COMPUTER SCIENCE EDUCATION RESEARCH

COMPUTER SCIENCE EDUCATION RESEARCH

EDITED BY
SALLY FINCHER AND MARIAN PETRE

Routledge
Taylor & Francis Group

LONDON AND NEW YORK

First published 2017 by Taylor & Francis Group plc, London, UK

2 Park Square, Milton Park, Abingdon, Oxfordshire OX14 4RN
52 Vanderbilt Avenue, New York, NY 10017

Routledge is an imprint of the Taylor & Francis Group, an informa business

First issued in paperback 2020

Cover illustration: *Pandora's box* by Christopher Pelley, with permission from International Arts & Artists, owner of *Tools As Art: The Hechinger Collection*, Washington DC, USA.

Library of Congress Cataloging-in-Publication Data

A Catalogue record for the book is available from the Library of Congress

ISBN 978-90-265-1969-7 (hbk)
ISBN 978-0-367-60453-0 (pbk)

Contents

Preface

Who Should Read This Book

This book is aimed at those new to the field of computer science education research, including postgraduate students (from a variety of backgrounds) and computer science educators who want to make the transition from a different disciplinary milieu. By providing a perspective on research in this domain, it may also be of interest to those already embarked on CS education research who wish to reflect on, or consolidate, their work. We also hope that it may be useful to others who would like an overview of the "state of the art" in this field.

We use the term "computer science" (and CS) throughout, through habit. We both spend our professional lives working with people who refer to themselves as computer scientists. However, we intend the term to be read in the widest meaning to include *Informatics*, *Computing, Information Systems, Software Engineering* and any other cognate term. This variety of terms, and names, can be taken as an indication of the immaturity of the discipline. Indeed, this immaturity is one of the motivations for the book which, among other things, attempts to provide a snapshot of "computer science" education research at the current time and current stage of development.

What This Book Is Not

This book is not comprehensive. It is not a manual of research methods, nor a tutorial, nor a "how-to" manual. Neither is it an atlas—it does not represent a completely understood territory.

Perspective, Orientation and Entry Points

We intend this book to provide an overview of how to approach CS education research from a pragmatic perspective. It tries to be as broad as possible, but we recognize that it cannot be comprehensive. So we have aimed to make it indicative of the range of motivations, of traditions, of research design, of techniques in an attempt to provide orientation for someone new to the neighborhood. We have aimed to represent the diversity of traditions and approaches inherent in this interdisciplinary area, whilst providing a structure within which to make sense of that diversity. In this way we have tried to provide multiple "entry points"—to literature, to methods, to topics—inviting readers to explore further for themselves, where their interests may take them.

The book is divided into two unequal parts containing two types of material. Part One *The Field and the Endeavor* is an introductory narrative that frames the nature and conduct of research in CS education as we see it. Part Two *Perspectives and Approaches* provides a number of grounded chapters on particular topics or themes, written by experts in each domain. They, too, aim to act as entry points, with illustrations drawn from published work.

Acknowledgements

Some of the material in Part One was developed from work supported by the National Science Foundation under Grant numbers DUE-0122560 and DUE-02443242.
We are grateful to Ian Utting, Peter Eastty, Josh Tenenberg, Andy Bernat and Simon Holland for intellectual and moral support throughout the long time it took for this book to come together. We are most grateful to the authors of the chapters in Part Two for their contributions (of course) but also for their enthusiasm, their professionalism and their patience. We also thank the *Bootstrappers* and the *Scaffolders* for participating in pilot studies for this material.

Sally Fincher, Marian Petre
Canterbury
November 2003

Part One: the field and the endeavor
Sally Fincher and Marian Petre

1

Mapping the Territory

Computer science (CS) education research is an emergent area and is still giving rise to a literature.

While scholarly and scientific publishing goes back to *Philosophical Transactions* (first published by the Royal Society in England in 1665), one of our oldest dedicated journals *Computer Science Education* was established less than two decades ago, in 1988.

Growth which has led to the emergence of CS education research as an identifiable area has come from various places. Some from sub-specialist areas: the *Empirical Study of Programming* (ESP) and *Psychology of Programming Interest Group* (PPIG) series of workshops; some from the major practitioner conferences which have included research papers—the Innovation and Technology in CS Education (ITiCSE) conference and the SIGCSE Symposium (now in its 36th year) run by the Special Interest Group in Computer Science Education of the Association for Computing Machinery (ACM).

Following from these starting points, there has been a burgeoning of publications appearing, perhaps opportunistically, in diverse locations, such as the IEEE *Frontiers in Education* (FiE) and the *American Society for Engineering Education* (ASEE) conferences, the ACM OOPSLA etc. Simultaneously, there have emerged a number of CS education research groups within academic institutions.

Despite this growth—and because of it—we are struggling to find the shape and culture of our literature. The task is difficult not only because the literature is distributed (there is no CS education *research* conference or publication) but also because our researchers and writers come from many established fields of scholarship and research—at least from education, psychology, computer science, technology and engineering. We have different intellectual traditions, and different conceptual frames, not to mention different methods and methodologies, and different reporting and citation styles.

What gets published from these different disciplinary areas, in diverse venues, are papers of very different types. However, in a simple-minded way, whatever their tradition, they can be thought of as having two components: a dimension of rationale, argumentation or "theory", and a dimension of empirical evidence. If we think of these dimensions as plotting a space, then four quadrants can be defined. On the top left, we have papers that have lots of argument, but little empirical evidence (although they may draw on other sorts of evidential material, similar to other disciplinary areas such as history). The bottom left quadrant *should* be empty; this is the home of papers with no evidence and no argument. The bottom right quadrant represents papers which are constructed around evidence—most often empirical— but are not strong on argumentation. Here is where descriptive, practice-based, "experience" papers are found, probably the most common type of paper in the area today. Finally, the top right quadrant represents papers that contain both evidence and argument. This is where, we contend, most CS education research papers *should* be found.

The concentration of papers in the lower, right-hand quadrant is representative of the state and status of CS education research today[1].

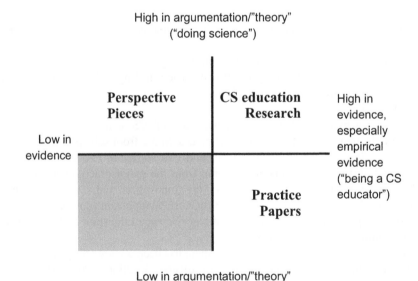

Figure 1: Diagrammatic representation of the types of paper found in CS education research
(after *Pasteur's Quadrant* (Stokes, 1997))

If we espouse a scientific approach to investigation in this area (and we do) then this includes the general trend and growth of scientific knowledge over decades. The process is, in general, an alternation of inductive reasoning (the discovery of general propositions) with deductive reasoning (the application of general propositions—once discovered—to particular cases considered to be included within their scope). Small-scale empirical studies, each designed to purposefully observe a phenomenon, build up to allow generalizations—"theories"—to be constructed. These generalizations can, in their turn, be applied in new situations. "These waves of alternate induction and deduction are superimposed, as it were, on a rising tide of which the general direction is inductive: a heaping up of a vast swell of generalizations which constitutes scientific knowledge as a whole" (Holmstrom, 1947). CS education research is not at a stage of development where many generalizations of this nature are possible: it is currently theory-scarce.

Motivating Areas for Computer Science Education Research

This section is principally a work of cartography. Its first aim is to map a territory of topics encountered in CS education research. Similar to all such attempts, there are some areas better known and better charted than others. Equally, this map will of necessity be historical, in that it must represent what people have found interesting enough to pursue to create a field. These topics and interests may not persist into the future, but taken together they should give a researcher new to the field a good idea of the current areas of investigation and interest.

We identify ten broad areas that motivate researchers in CS education. We use "motivate" intentionally here because we believe that these areas reflect key topics and highlight *why* researchers become interested in the first place. Each area has its own, sometimes small, tradition. Some we shall explore here, others are developed more fully in later chapters. Sometimes these areas, and their literatures, are encompassed within a single disciplinary genre; at other times they may have been examined from many approaches, and their only commonality is the interest inherent in the questions they pose or the nature of the phenomena they represent.

The ten areas are: *student understanding, animation/visualization/simulation systems, teaching methods, assessment, educational technology, the transfer of professional practice into the classroom, the incorporation of new development and new technologies into the classroom, transferring to remote teaching ("e-learning"), recruitment and retention of students*, and, finally, *the construction of the discipline* itself.

Student understanding
The area of student understanding is characterized by investigation of students' mental and conceptual models, their perceptions and misconceptions. The kinds of question that researchers find motivating in this area are concerned with *why* students have trouble with some of the things they have trouble with, what *distinguishes* good students from bad students, and what the differences are between how students understand things and how experts understand things. This area of interest encompasses investigations at a wide variety of scale from very broad

topics, such as "What design behaviors do students exhibit?" and "How do students learn in particular programming paradigms?" to very specific questions such as "How do students learn recursion?" Because of the nature of the evidence required in this area, given that it focuses on internal phenomena, some work has been conceived as CS education research, but there is also a lot of work of this kind that that occurs within psychology, or sociology. A good overview of some of this literature can be found in later chapters in Part Two, particularly *Research on Learning to Design Software* and *Misconceptions and Attitudes that Interfere with Learning to Program*.

Animation, visualization and simulation

There is a strand of research that draws on one of the peculiar facets of CS education research, which is that, as computer scientists, we can build things. We can devise and develop systems. We can do this to serve our research questions, to allow us to investigate and follow up conjectures. This has given rise to an area of research which uses software tools and environments to affect student learning. These tools broadly fall into the categories of animation (often algorithmic), visualization (often of processes within the machine) and simulation (of differing situations and conditions, traffic of differing network configurations, or simulating the results of using different search techniques). The kinds of question that researchers find motivating in this area concern the changes in teaching and learning when students can explore, enhance and even construct their own understandings. One of the unusual aspects is that these tools and systems are often the sorts of tool that interest educational researchers in *other* fields. It may be that, in this area, CS education research has a unique export. A good overview of literature in this area can be found in the chapter *Algorithm Visualization* in Part Two.

Teaching methods

It is clear that this is a very broad topic that could be broken down further. Broadly, there are two kinds of question that researchers find motivating in this area.

Firstly, there is work concerning how teachers can "build bridges" for students, how they can scaffold their students' learning, helping them to make sense of the subject. Sphorer and Soloway's work on programming plans (Sphorer, Soloway, & Pope, 1985) is an early example of work in this area, which has proved to be influential in the long term. Linn and Clancy, too, have had great impact with their work on case studies (Linn & Clancy, 1992).

Secondly, there is a body of work regarding how teachers control the dynamics of the teaching interaction to make it profitable. This often highlights activities and presentation methods constructed to advance student learning, e.g., (Astrachan, 1998; Astrachan, Wilkes, & Smith, 1997).

There is also some work that is theoretically motivated, building on findings from other disciplines (largely in psychology) with regard to issues like "active learning", "cognitive styles", and "learning styles". We discuss the basis for some of these ideas later in the *Link Research to Relevant Theory* section.

Other aspects of teaching methods research are those studies that take a single approach and trace it through many instantiations, sometimes many institutions. One such example is the Effective Projectwork in Computer Science project (EPCoS)

(Fincher, Petre, & Clark, 2001). This examined the place of projectwork in the computing curriculum, identifying different kinds of projectwork; how and where it occurs in the curriculum; and how teachers discover ideas about the topic, how they use it, and how they transfer it among themselves.

Assessment

Assessment is another broad area, which we break down in terms of categories of research question: types of assessment, validity of assessment, automated grading.

Some questions address distinctions among *types of assessment*, trying to understand which types are most suitable for particular assessment aims or contexts, and what makes them effective. Assessment may be formative (conducted at stages during the course of a project in order to contribute information that will influence, i.e., help to "form", subsequent stages) or summative (conducted at the end of a project in order to assess the project as a whole, i.e., to "sum it up"). It may address different kinds of learning, such as acquisition of factual knowledge, change in conceptual understanding, acquisition of skills. A good example is Lister and Leaney's work applying Bloom's taxonomy to CS education (Lister & Leaney, 2003).

Some research is aimed at understanding *whether the assessment is valid*, whether it represents the kinds of knowledge the educator wants it to assess. For example, results of conventional testing might be compared to conclusions of in-depth investigations of students understanding of the material tested, as was done at the Open University (Petre, Price, & Carswell, 1996).

Issues associated with *automating grading* form another strand. Among these issues are the assignment of partial credit, handling responses to open questions, verification that the examination has actually been taken by the named student, and plagiarism (e.g., (Lancaster & Culwin, 2004) and (Sheard, Dick, Markham, Macdonald, & Walsh, 2002)). Work in this area started very early. A more recent example is the Ceildh project (Foxley, 2003), a significant attempt to develop and research automated grading.

There are also studies that consider assessment in a broader context, examining assessment from a curricular or cross-institutional perspective. For example, one ITiCSE working group (McCracken et al., 2001) conducted a multi-institution study which examined what proportion of 'first competency' programmers could actually solve the sort of programming problems their teachers thought them capable of undertaking.

Educational technology

Many researchers find this area motivating, and not always for disciplinary reasons. Some of the work that occurs here is generic. In that way, it is similar to the work in visualization (above), drawing upon our disciplinary skills and interests, to investigate questions in the world. Just as every inventor can see opportunities to "build a better mousetrap", every computer scientist can see opportunities to build a better application. Some of the work in this area leverages the advantages that new devices and technologies offer. There is work on presentation systems, using tablet PCs (Anderson, Anderson, VanDeGrift, Wolfman, & Yashuhara, 2003), and Personal Digital Assistants (PDA) (B. A. Myers, 2001). There is also interest in whole environments, instrumented learning spaces, and "smart" classrooms (Abowd, 1999). There is also work being undertaken with (and using) computers as "assistive technology" teaching the visually impaired, or blind.

Of course, there are also examples of educational technology being applied to CS subject matter. Common examples are pedagogic environments for initial language learning, for example BlueJ (www.bluej.org) and DrScheme (www.drscheme.org). This area can involve hardware as well, for example the "smart" whiteboard *Ideogramic* for creating UML diagrams (www.ideogramic.com). A more detailed overview of the environments literature can be found in *Programming Environments for Novices* chapter in Part Two.

Transferring professional practice into the classroom
CS is a vocational discipline. Unlike some other academic disciplines, this means that there is a cadre of professionals who are developing and expanding (at least) the practices of the discipline, in parallel with academia.

One research strand takes as its focus the transfer of professional practice into the classroom. The motivations are clear. Academic researchers observe expert practice and say, "Our students need to know about that because they're going to have to make the transition into that environment" or, sometimes, "Perhaps if we taught it that way, our students' understanding would improve". Academics seek inspiration in the work of professionals, because some professional methodologies are specifically about good practice, about scaffolding people to "walk the walk" of software development in an efficient and effective way. A good example is XP, which typically manifests in the classroom as pair programming (Williams & Kessler, 2001).

Incorporating new developments and new technologies
Running alongside *transferring professional practice*, but separate from it, is an area which concerns itself with incorporating new developments and new technologies into the classroom. Some of the most ephemeral research in CS education research is in this domain, because industry moves fast. Educators see new developments and work to incorporate them. Sadly, only some of those developments last. For those that do, the research often ends up simply reporting on how to make a transition from old to new.

Transferring from campus-based teaching to distance education
Like all other disciplines, CS is increasingly taught in "non-traditional" settings. In the twenty-first century, "non-traditional" almost always means "at-a-distance-and-computer-mediated". Of course, there are many generic issues here about Web-based learning and appropriate transfer of educational interaction from a radically co-located environment to a remote one. An early pioneer in this field is the RUNESTONE project. Within RUNESTONE students work on a CS project (involving real-time systems) in teams, under academic supervision. The especially interesting feature of RUNESTONE is that half the students in each team are from Sweden and half are from the US (Hause, Almstrum, Last, & Woodroffe, 2001). They live and work in different time zones, and never meet face to face. Yet they

work on the same project, for which they are assessed as in any other similar piece of academic work.

Recruitment and retention

Issues of recruitment and retention motivate a research area, including a considerable interest, and body of work, in diversity and gender issues. There are real questions about what makes students come into CS and what makes students stay in CS. And the other side of the coin: what makes them not bother, and what makes them leave. Usually, there's a diversity perspective, but there doesn't have to be. It can be just one of the questions that concern the discipline. For example, what innate abilities contribute to performance in CS? What are the kinds of skills we can engender?

Construction of the discipline

The final category is of a different kind, concerning questions about the construction of the discipline. In some other domains, for example mathematics, there is a *didactics*, a sense of what it is we're supposed to teach, an acknowledgement of what we should cover as fundamental principles, and an associated understanding of which curricular areas are advanced and which are optional. There can also be a sense of how subjects should be taught, how they should be delivered. We clearly don't have agreement on that in CS, although the ACM computing curricula (http://www.acm.org/education/curricula.html) and, in particular CC2001 (http://www.computer.org/education/cc2001/final/index.htm), has generated a real discourse around these areas.

As well as curricula constructions, this area encompasses questions concerning the nature of the discipline: is it an engineering discipline? Is it mathematics? Is it design? Is it business? Is it something else altogether? And this leads to discussions of interpretation (Fincher, 1999) and scope; of how many things this discipline actually embraces.

What Computer Science Education Research Isn't

"Computer Science" is itself a new discipline, created out of many others. The debt to mathematics is clear, and often seen as the "core". Formal methods, theoretical CS, algorithms etc., all use mathematical methodology. Hardware interfacing is clearly akin to electronic engineering. Software engineering methodologies have come into academia from industrial practice; operational research and business processes have come from business schools.

CS education research is also informed by other disciplines. Much theory is from education and the learning sciences. Experimental and analysis techniques are drawn widely, from statistics to empirical studies to social science methodologies such as ethnography. In this way, the development of CS education research has been similar to many of the other research areas within CS which have developed from other subjects.

The key to viewing "CS education" as an area distinct from CS and from other disciplines must surely be in the questions we ask. If we ask questions that are generalist ("Do students learn better from face-to-face or Web-based interaction") or facile ("Do students learn better if A or B is their first programming language"), then perhaps practice descriptions may be written, perhaps more pedagogic environments and visualization tools may be built, but perhaps what we are doing (although valuable) is not CS education research.

But if we ask questions that can *only* be addressed from within CS, perhaps "Does a knowledge of computer architecture make better—more expert— programmers?" or "Does student understanding of programming concepts differ with language of first instruction?" then these questions *cannot* be addressed by someone outside of CS. What meaning, after all, would such questions have to someone who could not program? Who did not understand the quality of the relationship between, say, functional and object-oriented programming and its import for first instruction?

There is a lot of material published in the area of "computer science education", and much of it is very good. However, the majority of it is not computer science education *research*, and the ways in which "computer science education" papers are good and bad are different from the ways in which "computer science education research" papers are good and bad.

CS education research is new. It co-exists in places with other sorts of publication (like SIGCSE), and where it starts and stops, where the edges of the endeavor are, is not yet entirely clear.

2

A Preface to Pragmatics

Science is not a cut-and-dried body of knowledge which someone has collected once and for all: it is an attitude of mind, a way of finding out. Unless these facts are appreciated science degenerates into mere scholarship and its study has a narrowing instead of a broadening effect on the mind. (Holmstrom, 1947)

Science is a discourse. The "stuff of science" resides not solely in data and variables, in hypothesis and observation—but in ideas and reasoning, in reflection and critique, and in conversational interactions among scientists who aspire to explain the world.

Method of Science vs. Scientific Method

At a workshop I attended not long ago, my colleague Matthew Chalmers made the observation that computer science is based entirely on philosophy of the pre-1930s. Computer-science in practice involves reducing high-level behaviors to low-level, mechanical explanations, formalizing them through pure scientific rationality; in this, computer science reveals its history as part of a positivist, reductionist tradition. (Dourish, 2001)

We're familiar with thinking about science in terms of "scientific method", an approach to theory validation based on Karl Popper's (Popper, 1959) hypothetico-deductivism. Popperian science proceeds by refuting hypotheses. Hypotheses are specific operational predictions that can be tested *empirically*. He distinguished science from "pseudo-science" by the notion of falsifiability, the doctrine that hypotheses are tested in order to demonstrate that they are wrong.

Experimenters can (and did) say "as impossible as a black swan"—for, in their world, all swans were white. Because one has only seen white swans, does that mean all swans are white—and that they are *all* white, forever? Well, maybe. But if you see just one black swan, then the theory is disproved. It doesn't matter how many white swans you see before that or after that one black swan, it doesn't make the theory any stronger, or truer. In other words we can't have true theories, only theories that haven't yet been disproved.

Common thinking has adopted a "shorthand science"[2], shifting the focus from the *meaning* of rigor in science to an *adherence to procedure*: as if following a set of rules will produce the desired result. The shorthand is that "science" is seeking knowledge through scientific method. By viewing science as a discourse, we need to put scientific method into perspective. Precise, controlled experiments are *one* approach to rigor, but not the only one, certainly with regard to the human sciences. There is no such thing as a "blank state human", a person without experiences, abilities, and knowledge. Hence we cannot control all human variables, and a strict experimental approach can exclude large portions of human experience. Human behavior—governed by fallible human perception and variable human cognition, and occurring in complex social contexts—presents problems for the pursuit of a straightforward one-to-one relationship between things and events in the outside world and people's knowledge of them. The complexity of human social reality makes it impossible to establish "facts" about behavior unequivocally. Falsification allows us to eliminate some theories, but we can't be sure that what survives is a "true" theory. We work with "best approximations", in which we have more or less confidence, depending on what evidence supports them.

In research involving human beings, it can be useful to think more broadly, in terms of a "method of science", characterized by principles such as articulation, validation, exposure to falsification, and generalization. "Method of science" demands rigor and seeks to contribute to empirically-founded theory, but it does not view "scientific method" as the only way of gaining knowledge. Rather, it seeks information using a variety of methods. This "method of science" admits a broader approach to rigor, and sees the construction and discussion of theory as closer to the ideas of Thomas Kuhn (Kuhn, 1970), who described science as a social endeavor which passed through a series of "paradigms", each paradigm persisting until a more comprehensive (and popular) explanation supplants it: the canonical example is the way that (in Physics) Einstein's relativistic paradigm supplanted the Newtonian one.

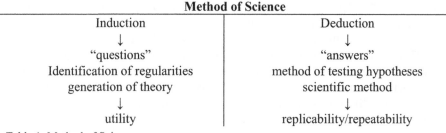

Method of Science	
Induction	Deduction
↓	↓
"questions"	"answers"
Identification of regularities	method of testing hypotheses
generation of theory	scientific method
↓	↓
utility	replicability/repeatability

Table 1: Method of Science

"Method of science" values description as well as hypothesis generation. It sees the relationship between theory and evidence as two-way, allowing both theory-driven and data-driven investigations. As shown in the *Method of Science* figure, above, "method of science" embraces both inductive and deductive reasoning. Deduction tests the predictions or hypotheses derived from a theory in ways that allow them to be disproved; induction draws inferences from observations in order to make generalizations. "Shorthand science" emphasizes deductive approaches at the expense of inductive approaches, but they *both* have value and *both* contribute to the discourse.

A combined approach can strengthen an investigation and make it less likely that critical factors will be overlooked, particularly in investigations of complex systems. For example, using induction to observe patterns in programmers' use of representations (e.g. graphical *vs* textual; data-flow based *vs* control-flow based)— and also in people's use of representations in other contexts such as cartography and music—Thomas Green generated a "match-mismatch" conjecture about the influence of representation on task performance. Using deduction, he predicted that people using representations would perform best when the information made accessible by the representation matched that required by the task. He and others tested the conjecture empirically by measuring programmers' performance (Gilmore & Green, 1984).

At the heart of the distinction between "scientific method" and "method of science" is the role of theory. Scientific method uses theory as a driver and a goal: theory generates hypotheses that can be falsified, and scientific enquiry produces predictive theory. With "method of science" what we are seeking is not formalism *per se*, but articulation. "Method of science" recognizes that the process of articulation is a crucial part of science: making explicit in sufficient explanatory detail. Articulation is needed for assumptions, meanings, constructs and their relationships, in order to help clarify the contribution to the discourse.

Epistemology

We need to be aware of what we understand to be the substance of science, of what we believe constitutes "knowledge", what we mean by "truth", and what we allow as "theory". In other words, we need to be aware of epistemology, of the philosophical stances underlying our research. Many different words and phrases are used in this area. Some are: *theory, explanatory theory, empirical laws, models* (sometimes "*explanatory models*"), and *conceptual frameworks* (sometimes "*conceptual lenses*"). Exploring what they might mean for CS education research and how we can use them to strengthen our work helps frame our endeavor.

"Theory" can be a problematic term for researchers whose background is in analytic disciplines. For the natural sciences, "theory" means "understanding that will generalize across situations or events", and it necessarily has a predictive and causal quality. "If I do *this*, then *that* will happen". Often it is bound up with notions of hypothesis, experimental setup and scientific method, and with representations that are not only precise, but also capable of formal expression—often mathematical. "Theory" in this sense works extremely well, and is responsible for many of the great advances of the last few centuries. However, it relies on certain fundamental circumstances:

- it relies on an unchanging natural world (the speed of light is always the same in Brisbane as it is in Berlin);
- it relies on accepted experimental procedures (external variability can be controlled so that what the experimenter observes actually is caused by what they think is the cause and not some extraneous, unaccounted factor[3]), and
- it relies on accepted methods of analysis, so that the "truth" of a knowledge claim can be assessed (you can/can't tell that from Bayesian analysis).

CS education researchers who start from this perspective, often flounder and "give up before they begin" in part because of the difference in research conditions. Learning environments are continually changing and hard to control.   In CS education research we do not have an established "common ground" of methods of investigation and analysis. Naïve researchers either try to tie down the world to a ridiculous extent in the hope that their experience will generalize (pre- and post-tests; teaching two sections with the same teacher, same aims but different materials/approaches) or give insufficient (or the wrong sort of) evidence to convince: they describe inessentials of the course setting in detail and then say "the students seem to like it". Considering a separate notion of "theory" can alleviate some of these problems.

> To explain the phenomena in the world of our experience, to answer the question 'why?' rather than only the question 'what?' is one of the foremost objectives of empirical science. (Hempel & Oppenheim, 1965)

Explanatory theories (also called "mid-range" theories)
Explanatory theories are characteristic of knowledge in the social and human sciences. The purpose of an "explanatory theory" is to explain observed human behavior (i.e., it is not predictive)—although generalized theoretical understanding is still a goal. The human sciences seek to encompass the complexity of human systems, actions, and experiences, understanding them as influenced not just by what can be observed and measured but also by intentions, motives, attitudes, and beliefs. Explanatory theories may not seek to account for cause and effect, but rather

to tease out and explicate factors and conventions that mediate human action and understanding in particular situations.

Those with an analytic background often deny that there can be any truth or utility in such an approach, but partly this is an under-representation of the word "explanatory". Work in the human sciences is often based on small numbers of very specific situations that are explored in a "deep" fashion. That is, many aspects are explored in many layers (explaining the Second World War, for example, requires many different layers of explanation, which take into account many different influences and affects; detailing the cultural norms of "gift-giving" require understanding of anthropological theory, specific cultural understanding and the vision to relate the two; the question "Do prisons work?" necessitates a mix of political, economic and psychological approaches, together with their respective theory bases.). However "theory" in this context is still *outside* of the circumstances in which observation/evaluation/experimentation take place. It is external, added-value intellectual effort that the researcher brings to the work. All too often, however, in CS education research, this is quite absent: the work is simply descriptive and in no way explanatory.

So, a "theory" in the natural sciences is concerned with cause-effect and predictability; a "theory" in the social sciences is concerned with the reasons for effects, the causes of which may not be determined, or determinable.

Empirical laws
Given these intellectual contradictions, many CS education researchers strive to generalize to "empirical laws". These are well-understood, simple quantitative predictions of human performance. *Empirical laws* gain their validity from statistical significance derived from sampling large populations. They may or may not apply to any given individual.

Largely, for our purposes, these are derived from cognitive psychology. Limits on short-term memory, for example, clearly have implications for complex cognitive tasks (such as programming) and on the design of systems to support such activities. Cognitive psychology has demonstrated that most human beings have a short-term memory capacity of seven, plus or minus two, things (Miller, 1956). Other disciplines also strive to discover (and utilize) empirical laws. There is no doubt that much medical research is devoted to discovering how the "average" human will respond to a specific intervention (normally drugs). Like "plus or minus two" this data comes with error boundaries—hence the list of "possible side effects" that comes with every bottle of aspirin. Epidemiology, too, relies on most people being "the same", so if there is a peak in, say, infant deaths in Canterbury in 1597 then it can be taken as an indicator of something out of the ordinary. Educational psychology also accrues data of this kind. SATs, IQ tests and other forms of standardized testing rely on (a) the quantity of people who take the tests and (b) the fact that what is tested is an ability ("intelligence") that every person possesses in some measure.[4]

Models
A common feature of theory and theoretical work, of whatever nature, is the formation of models. Models are generalized, hypothetical descriptions of

something that is not directly observable. Sometimes, the generation of a model is a sufficient end of a theoretically-inspired investigation. Mostly, it occurs as part of the process of research.

> Models are inevitably simplified versions of reality. They can only aim to select the more important variables in a complex problem, so as to provide a manageable insight into the issues. (Kember, 1995)

Very often models are based on an analogy: an atom is a model (the most common analogy is to the solar system, with the electrons "orbiting" the nucleus); culture is a model. In every analogy some things are the same, and some things are different. It is worth thinking about what is important to have as similar, and what can safely be different. Some researchers consider models to be "second best". If we can explain things by cause-and-effect, that's best; otherwise we'll make do with comparisons and analogies.

Conceptual frameworks (or "conceptual lenses")
A conceptual framework defines a particular point of view within a discipline from which the researcher focuses his or her study. This "theoretical perspective" identifies underlying assumptions from which particular kinds of questions are generated.

For example, when researching ecological impact and sustainability there are a variety of conceptual frameworks within which a researcher can work, including "net primary productivity" (NPP) and the "ecological footprint" framework (OECD, 1995).

NPP is the amount of energy left after subtracting the respiration of primary producers (mostly plants) from the total amount of energy (mostly solar) that is fixed biologically. The researcher who takes an NPP framework also takes the underlying assumptions that human co-option of terrestrial resources contributes to the extinction of species and will shut out a number of options for humanity. Consequently the measures (or indicators) that they use are constrained by the framework and its assumptions.

Ecological footprints are calculated by a population's demand for domestic food, forest products and fossil energy consumption, converted into the required area of eco-productive (agricultural and forested) land. An ecological footprint provides an area-based indicator of the physical limits to material growth. The researcher who adopts an "ecological footprint" framework takes on the key assumptions that industrial economies currently survive through importing the "surplus" carrying capacity of developing countries. This pattern of consumption activity implies (1) that developing countries are restricted in their own development (insufficient carrying capacity available) and (2) that developing countries' desire to emulate Western living standards cannot be fulfilled since there is insufficient global carrying capacity: the footprint already covers the earth.

So, in this example, we see that both researchers are interested in human/environment interaction, but, depending on their conceptual frameworks, they ask different questions and use different methods and indicators to provide evidence to answer those questions.

Relationship between phenomena and evidence: two examples
It is clear that much of science is concerned with the observation of circumstance, the recording of phenomena. However, what is the nature of observation? What does it mean to look at (or for) something particular? For something that will support (or disprove) what we believe to be the case? What is it that turns the *observation of a phenomenon* into *evidence that supports a theory* or theoretical approach? To illustrate the problems of phenomena and evidence we take two examples, one from physics and one from medical history.

Physics
Since 1910, physicists have (with increasing sophistication) been able to record the "trails" made by sub-atomic particles in cloud chambers. Sub-atomic particles had been *theoretically* postulated previously, of course, and there were *models* of what atoms were and what they were made up of. But in 1910 photographs became available for the first time. What was the epistemological status of these pictures? Are they simply recording a phenomenon that happens in the world anyway, whether we can directly see it or not? Or do those trails provide evidence of the very existence of those particles, which were only previously theorized about?

There were many opinions expressed about the pictures, from "I understand your lingering doubts ... I think you would be convinced if you looked at the photos" through "I myself was fairly convinced of the reality of the phenomena though naturally other possible interpretations were not inconceivable" to "Any ... record would be misleading which did not adequately stress the element of critical interpretation which is necessary" (all reported in (Galison, 1997)). The photos lay precisely on the point of tension between what theory predicted and what experimenters saw. The relationship between the record of a phenomenon and evidence that supports (or denies) a theoretical position, is *interpretation*.

Medical history
Edward Jenner (1749-1823) was a doctor in a rural English setting. He noticed that, within the community he served, people who worked closely with cattle and had caught a mild infection from them (called cowpox) did not get smallpox—at that time a disfiguring, often deadly, and much-feared disease. He found this curious, and he compiled extensive case studies and observations to find out if this was, in fact, the case (Jenner, 1798). He found that this was so, and went on to invent the process of "vaccination" protecting people from smallpox.

But what is the epistemological status of his case studies? Are they simply recording a *phenomenon* that happens in the world anyway, whether we can see it or not? ("Oh yes, Sarah had cowpox and didn't get smallpox.") Or do those compilations provide evidence of the very existence of something only previously theorized about? ("There is a relationship, not between specific people and circumstances, but between these diseases.")

CS education research
Much of what is published as CS education (called "research" or not) has been concerned with *noticing phenomena*: "This is what happens in my classroom", "This

is what happens when you teach *x* in this way", "If I teach *x* differently, something else happens". What moves recognition of phenomena to *evidence* is purposeful investigation and a relationship of investigation to theory.

Truth claims
Underlying these issues—of epistemology, of theory, of phenomenon and evidence—is how we can know something is "true" and how we share that knowledge. Different parts of life have different systems of creating and understanding knowledge, which are not transferable to other systems. For example, religious truth-claims find their validity in the epistemology of religions, and judicial truth-claims find their validity in the epistemology of the judicial system, but we can't apply the way we believe in a God to the way we judge a criminal. The evidence is different, the "burden of proof" is different, and the way we share the results with others is different.

Scientific truth-claims find their validity in the epistemology of "science" which we in our consideration of "method of science" take to be meaning-filled public discourse regarding replicable experience. Validity (or "truth") in the discourse of science involves a combination of factors, including the accuracy of our observation, the quality of our reasoning, and the completeness of our explanation.

The kinds of research questions that can be asked are (partly) dictated by the researcher's epistemology. This underpins (and constrains) what *kinds* of question can be asked, what are *legitimate* questions, what are *appropriate* questions, and even what questions are *allowable*. Epistemology provides ways of conceiving and seeking knowledge which focus endeavor. Theories and models provide reasoning frameworks, which highlight important relationships, focus enquiry and hence simplify.

Pragmatism: striking a balance
Our approach to CS education research is pragmatic: demanding rigor, aspiring to generating fact and theory, but accepting "best fit" *en route*, and using whatever research methods contribute to the process and discourse. As pragmatists, we needn't argue about whether we can derive general, predictive models for CS education or we can only explicate specific evidence – we can use explications as a step toward explanatory theory preliminary to predictive theory, and not worry about whether the predictive theory will actually be achieved. We strive constantly for rigor, but we seek rigor *in terms of* whichever approach we're taking.

Each discipline has its own rigor and standards; each study must be rigorous in its own terms. It is essential to understand the underlying premises of any approach employed. Importing methods alone is insufficient—the researcher must understand the knowledge concerns, and the associated assumptions and limitations, not just the application of techniques. There can be multiple ways of studying a phenomenon, but that doesn't mean "anything goes". The conceptual framing of each method must be understood and respected. This is especially true if different approaches are to be combined or compared. This is at the heart of the "method of science". As Liam Bannon said, "Science is not just a set of methods, but a way of reasoning."(Bannon, 2000)

The CS Education Research Endeavor

Our approach is pragmatic. We do not advocate research driven by any one theory, discipline, or methodology. Rather, what we intend to do in the next sections is outline the necessary process for conducting effective empirical research in CS education. We concentrate on what to do, not how to do it.

We shall take as a framework for the next sections, "six guiding principles" for research that have been formulated by the *Committee on Scientific Principles for Education Research* and which were published in their book *Scientific Research in Education* (Shavelson & Towne, 2002) . There are several reasons why we do this: Firstly, the principles they have identified are good ones. Secondly, they are formulated at a sufficient level of abstraction so that they are not bound to any specific research stance, or tradition. Thirdly, they embody "method of science".

In each section we link these principles to CS education research, its nature and status, and illustrate some of the problems that are bound with research in CS education.

The six principles

- Pose significant questions that can be answered empirically
- Link research to relevant theory
- Use methods that permit direct investigation of the question
- Provide a coherent and explicit chain of reasoning
- Replicate and generalize across studies
- Disclose research to encourage professional scrutiny and critique

The principles are not always equally weighted. Depending on the nature and quality of the investigation, they may have different significance and importance. Also, they may be considered in a different order for different studies.

3

Pose Significant Questions that Can Be Answered Empirically

What we mean to offer is a pragmatic approach to empirical research in CS education, because CS education research will necessarily involve compromises. The pragmatic approach centers on understanding the value of evidence and its fitness for purpose—its utility.

Designing Empirical Studies: 1 – 2 – 3

Our advice about how to design empirical studies for CS education research always follows the same formula:

1. Figure out what the question is
Figuring out what the question is, is probably the most intensive step. It requires identifying what is important for you to know, out of what might be known. It requires an assessment of whether what you want to know is something that can feasibly and reasonably be investigated. It typically involves an analysis of whatever motivated you to ask the question in the first place, and it often involves resolving an initial question into a smaller, more tractable questions.

2. Decide what sort of evidence will satisfy you

The next step is deciding what sort of evidence will satisfy you—actually, will satisfy a reasonable, skeptical colleague—*in addressing or answering the question.* What would a sufficient answer look like? Determining the evidence requirements involves figuring out how the phenomenon of interest might be manifest in the world and hence how to 'operationalize' your question, to phrase it in terms of things you can observe directly. It also involves learning the value of different types of evidence and assessing how strong the evidence must be to serve your purpose.

only then…

3. Choose a technique that will produce the required evidence

In this pragmatic approach, methods or techniques follow from the question and the evidence requirement. This is equally true of theory-driven and inductive studies.

Researchers who start by asking *how can I design an experiment in order to prove X* are starting from the wrong place, because they've chosen a technique before they've really considered what the question is. Even well-designed experiments are of little utility if they address the wrong question—and it's difficult to design an experiment well if the question is not clearly in focus. Good experiment design requires an understanding of the key variables, as well as a precise question.

For example, consider: "How can I design an experiment to prove that my students write better programs when they use functional programming?" This may seem clumsy now, but not long ago it was a familiar query. However, it's packed with assumptions and floundering in imprecision. What is a "better program"? Why might you think that functional programming is a solution? What is it a solution to? On the reflection, there is a fundamental question underneath—one that must be addressed even before we can begin to consider the role of any particular language (or treatment): "What is the underlying need or deficiency which a shift to functional programming is trying to address?" Suddenly the question is not phrased in terms of "the application of a treatment" and "proof", but rather in terms of teasing out factors that are not yet identified and articulated. And suddenly it's evident that the question is not at the level of *testing* an identified relationship, but of *seeking insight* that might lead to the identification and explication of relationships.

Premature commitment to experimentation is like using a sharp scalpel when a chainsaw is more appropriate. Much of what CS education researchers want is at the chainsaw level of "finding better questions" and of generating early theory. Once the questions are brought into focus, then a scalpel-level experimental regime can help in dissecting details, in testing emergent theory. Experimentation assumes theory that we don't have. Experiments can only *dis*-prove. Prioritization of the question reveals what sort of information is really needed, and makes clear that techniques must follow questions.

Operationalization

At the heart of question formulation is *operationalization*, a process of mapping from what we want to know to what we can investigate empirically, that is, to what we can observe in the world. We need to link the concept or construct of interest onto one or more manifestations in the world, things that can be observed, recorded, and ultimately measured in some way. The validity of the study rests on that operationalization, on that mapping from construct to observable phenomenon to measure. Researchers are often distracted from what *should* be observed by what *can* readily be observed. If the reasoning that associates the manifestation with the construct is faulty, then the data may be irrelevant or misleading.

Operationalization is typically a simplification. An operationalization might be selective, might focus on one facet of a phenomenon. An operationalization might be a shorthand, a compact expression or reflection of a phenomenon, for example a measure. But the things we capture—recordings, descriptions, categorizations, measures—are *not* the phenomenon.

Operationalization is important not just to our conduct of a valid study, but also to the value of the study in the discourse. We must be able to communicate the operationalization, in order for one operationalization of a construct to be compared to another, in order for the findings of one study to be related to another, in order for evidence to accumulate across a number of studies. Not only is the construct mapped onto some manifestation in the world, but also the mappings applied in different studies must be compared. Is the construct interpreted in the same way? Are the manifestations comparable? Are the measures applied to the manifestations actually measuring the same thing? How well do the measures reflect the manifestation, and how well does the manifestation represent the construct?

Sorting Out What the Question Is

A question well-stated is a question half-answered. (Isaac & Michael, 1989)

Formulating the question well sets up the rest of the study design. Some questions arise from theory, some from observation or from conjecture based on observation. Question formulation is itself a crucial reasoning skill. There is a refinement path from an initial, intuitive question to a well-specified question worth asking, a question that relates to what is already known and understood, is significant, and can be investigated.

Why ask?

A good first step in formulating a research question from an intuitive curiosity is to consider what kind of evidence made you think the question was worth asking in the first place. The question is not just "why ask?", but "why bother asking further?" Considering both why that evidence (be it introspection, anecdote, a classroom observation, a line of discussion in the literature) was enough to make you ask—and what is missing from that evidence in providing an answer—can provide insight into specifying the question.

What would not suffice?

A good second step is to consider what sort of answer would be inadequate—what a "non-answer" would look like. This can help to clarify the evidence requirements. It can also help in distinguishing the question of interest from other, related questions.

Counter-examples?

A good third step is to consider what a counter-example or a contradiction might look like. How might the conjecture be falsified? People tend to seek confirmatory evidence, to try to demonstrate what they believe to be true. But often insight lies in the "surprises", the unexpected, the contradictions. Considering the nature of counter-evidence can be a way of reflecting on the basis of the question: the observation, conjecture, or hypothesis underlying it. It can be a way of exposing alternative accounts, which in turn can be a way of exposing inadequacies in the question formulation.

Work back from the analysis

Another useful approach is to think "backwards", to consider what relevant data might look like and how it might be analyzed. Plan the analysis with the study.

So what?

One of the tests of a worthwhile question is "so what?", or "what will I do with the answer if I get it?" In formulating a research question, it's important to consider the question not just in its own terms, but also in the context of the discourse to which the study might contribute. Hence, it will be important to establish the:

- *importance* of the question
- *significance* of the findings
- *implications* for theory
- *limitations* to generalization

These are appropriate considerations when framing the research question, because they distinguish questions worth asking.

Questions to avoid:

It's important to identify questions that are worth answering. Which means that some questions are better avoided:

- *the unanswerable:* There are questions that, although compelling, are too abstract, too elusive to operationalise, or simply far too costly to address effectively.
- *overworked topics:* Some questions have received considerable attention from others, limiting the likely impact of any one study. (e.g., "Can mathematics grades be used to predict success in CS1?") Unless a researcher can bring surprising novelty and insight to an overworked topic, or provide a definitive study, the topic is best avoided. Sometimes, a parallel inquiry in a different

context or addressing a different but structurally similar area can unlock an overworked topic.

- *trivial topics:* These are questions whose scope is too small to make them generalizeable, relevant, or interesting to others. (e.g., "Do colored mice improve student productivity?")
- *"one-shots":* These are questions that don't aggregate, that don't contribute to an accumulation of evidence.

Table 3, below, gives a light-hearted summary of some of the most common pitfalls in empirical study design. Most of them can be addressed at the planning stage, during the cycle of reflection that focuses the question and identifies its evidence requirements.

Evidence

The key to empirical studies is in knowing the value of *evidence*. Evidence is at the heart of the scientific discourse. Theories are supported or contradicted on the basis of evidence. Rather than thinking of studies as "good" or "bad", it's more productive to think of them as producing "strong evidence" or "weak evidence". Whether or not a study is "good" depends on whether it produces sufficiently sound and convincing evidence *for its purpose*.

Some purposes need only weak evidence or contrary examples. For example, a researcher might be content to make a minor interface design decision based on the performance of a handful of subjects on a simple task. When one is trying to dispute a universal claim, one only needs a single counter-example, one "black swan". Other purposes require strong evidence. For example, deciding whether to re-vamp an entire curriculum around project work suggests a substantial investment and probably demands compelling evidence of efficacy.

In CS education we can't hold enough of the world stationary to achieve precise control. "Proof", therefore, is impossible. Rather, the discourse examines the relative strengths of competing claims in terms of the evidence that supports them—and the evidence that contradicts them. So, we must decide what matters and how it can be investigated, whether by counting or through a qualitative method; i.e., we must understand the value of evidence.

A second key is to remember that research studies do not stand-alone; they must be assessed in the context of other studies that provide relevant evidence. Results accumulate, and studies may be repeated with different subject groups or with slight variations in order to explore the reliability (consistency across repetitions) and robustness of the findings (consistency across slightly different settings). Different studies may combine methods, or "triangulate", in order to overcome the shortcomings of any one method.

Research is about *learning* (i.e., adding to knowledge), not proving. Discourse and scrutiny are as important as outcomes in developing theory. The purpose of empirical research is not only to observe behavior, but to *think about* behavior. Empirical science in young domains such as CS education is not so much a process of getting answers as one of finding ever better questions. We are unlikely to achieve total accuracy; total generalizeability; realistic integration, comprehensiveness, or completeness. But *so what*? Will we be able to ask better questions?

Data is not necessarily evidence

> There is something fascinating about science. One gets such wholesale returns of conjecture out of such a trifling investment in fact. (Mark Twain)

Data is not necessarily evidence. Data *becomes* evidence when its relevance to the conjectures being considered is established (Mislevy, 2001). The path from data to evidence is *inference*—reasoning about what is observed and what it means. Inference is reasoning from what we know and observe to conclusions, explications and, possibly, predictions. The value of evidence relies both on the quality of the *data* (deriving from the appropriateness of its selection, the quality of its capture, and its representativeness) and on the quality of the *inference* that connects the data to the phenomenon of interest.

Knowing the value of evidence:
quality of data + quality of inference

Evidence is not proof

Evidence is not proof. In general, it is whatever empirical data is sufficient to cause us to conclude that one account is more probably true than not, or is probably more true than another.

In order to substantiate evidence, we must establish its:

- plausibility (is it likely, given existing knowledge)
- validity (that it is a true reflection of the phenomenon under investigation)
- relevance (the data relates to the research question)
- credibility (whether the researcher's interpretation is likely to be accurate)
- inferential force (the legitimacy of the chain of reasoning).

We need to understand the value of different forms of evidence and how they fit together. We need to understand how reliable the evidence is likely to be (how consistent the outcomes will be given repetition by different researchers, at different times, with a different sample of the same population), how robust (how consistent the outcomes will be across different related tasks, across different environments, across different related contexts), what margin of error it entails. In the same way that we report the standard deviation associated with a mean, we must report the uncertainty and error associated with evidence—hence enabling assessment of its fitness for purpose.

Richness and rigor

We need to strike a balance between the richness of realism and the precision of controlled studies, exploiting the advantages and compensating for the disadvantages of each. As Bill Curtis has said, "We need experience in a real environment to figure out what the critical variables are…practical experience leads to a better development of relevant theory".

Table 2: Some characteristics of evidence: richness and rigor (Informed by ideas from Marilyn Mantei, D.A. Schkade, and Bill Curtis)

Richness/ Realism	Rigor/ Control
Reflecting reality (Natural phenomena) 'Practical' Valid	Replicable Repeatable Reliable
BUT	
"Real data is real dirty."	Sand through the fingers

'Richness' indicates the number and generality of questions we can attempt to address using the available evidence. The very process of squeezing reality into the requisite structure of a controlled experiment often strips it of some of its worldly richness—and hence constrains the scope of the question it can answer. Equally, the very richness of realistic investigation makes it hard to draw conclusions that are not context dependent, and possibly specific to the particular instance.

So, on the one hand designing experiments in a CS education context can be like "sand through the fingers": we try to grasp an issue, but it slips away as we sift through factors in a search for control. On the other, as Bill Curtis so eloquently phrased it: "Real data is real dirty." In effect, realism trades off with control, richness with precision.

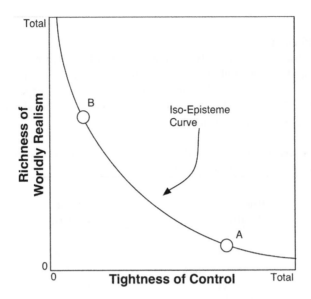

**Knowledge Domain Defined by
Theory and Conceptual Variables**

Figure 2: Richard Mason (Mason, 1989) has characterized the richness/rigor tradeoff as a
space populated by iso-episteme curves, each curve representing points of equal
knowledge generated by various forms of investigation. In the figure, points A and B
represent equal knowledge, but they are qualitatively different. A might represent a tightly
controlled experiment, whereas B represents a study that incorporates considerable reality.
Mason notes that the amount of knowledge generated depends on the skill and care of
study design and execution; a well-conducted study would be placed on a higher curve
than a sloppy one.

Richard Mason (1989) has characterized the richness/rigor tradeoff as a space
populated by iso-episteme curves "which represent the fundamental tradeoffs that
must be made between these dimensions in conducting research", as shown in
Figure 2. His analysis, too, is pragmatic. He observes that all empirical studies can
be improved, but that improvements are made at a cost.

Granularity and focus are important concepts here. The granularity of the
evidence must relate to the focus of the research question. Very precise studies may
leave too much of the question unaddressed; very realistic studies may illuminate
too many factors to provide insight about the particular question—in either case, the
mismatch between study and question mean that the resultant evidence does not
serve its purpose adequately.

There are many forms of evidence we tend to think of as weak. A good example
is "anecdote", which is often used to motivate studies, but is rarely accepted as
sufficient evidence. But a *collection* of anecdotes, gathered from a variety of
independent sources, can provide much stronger, more convincing evidence (as in
Eisenstadt's study of "tales of debugging" (Eisenstadt, 1993), sufficient to refocus

research questions and provide a basis for comparison to other forms of evidence. The point is that purpose (what are the findings to be used for?) gives insight into pragmatics (what sort of evidence is sufficient).

Our perspective is that we need *both* realism and control in our studies of human behavior and reasoning, in our case the behavior and reasoning of students learning computer science, but not always in the same study. Concept formation and definition are a key part of science, and of the scientific discourse. They are preliminary to the operational definition needed for experimentation. Our focus is not on techniques per se, but on the accumulation of evidence fit for our purposes.

Triangulation and multi-method research

A way to strike a balance between realism and control is through the use of multiple methods over a series of studies that accumulate evidence. *Triangulation* is a name for combining techniques (for example in a series of studies of differing design) in order to shed light on an issue from different perspectives and thus overcome the limitations of any single technique. It can also be a method of challenging or corroborating findings, by gathering and comparing multiple data sources against each other.

Multi-method research is a way of both increasing the richness and increasing the rigor, sometimes within a study, but more often across a series of studies, e.g.:

- multiple factor studies (e.g., system design and task)
- multiple "dependent variables" (e.g., not just performance, but change of performance with experience) within experimental and constrained task designs
- multiple measurement methods
- multiple research methods

Of course multiple-method research doesn't come "for free". Each of the methods must be understood in its own terms: the assumptions that attach to it, the level of resolution of the data it generates, its constraints and limitations, and so on. Its appropriateness in generating data relevant to the research question must be considered; the method must be fit for its purpose. The consequences and challenges of combining methods must be understood, lest inappropriate comparisons or faulty inferences be made. The application of each method must be valid, and the analysis and interpretation that links them must be sound.

Multi-method research carries a burden of responsibility: we must take care that when we 'connect the dots' we do so legitimately. This involves:

- Respecting the epistemology and traditions that influence (and constrain) the methods we borrow.
- Articulating the reasoning by which we establish associations and make comparisons among disparate data, and taking care when drawing inferences beyond the granularity and focus of a given study and its evidence.
- Identifying constructs and developing operational definitions that are valid.
- Explicitly considering constraints, threats to validity, and possible alternative accounts.

- Confining conclusions to what is supported directly by the data, noting the limitations of our results, and taking care in recognizing when generalization is conjecture only.
- Indicating the level of uncertainty and error associated with the evidence.

Multi-method research can increase rigor through triangulation, both by providing opportunities to challenge findings and to expose alternative accounts and by the accumulation of evidence, using different techniques to provide different approaches to a phenomenon. Pragmatically, rigor comes from using the best procedures we can devise to make the *best effort* to avoid error (such as observing principles of repetition, exposure to falsification) in order to get the best information/knowledge that we currently can.

Utility
At the heart of this perspective on evidence is a concept of "utility", that the "goodness" of evidence relates not to some absolute standard, but to its "fitness for purpose", to its relevance and strength in terms of the use to which it is to be put, to its usefulness for understanding some phenomenon of interest. The utility of evidence is what establishes its value.

Utility applies throughout the empirical chain, from determining the focus and granularity of the question, through the collection of data, through the analysis. Each stage must match the evidence requirement, be it strong or weak. Sometimes there is a need for definitive, generalizeable answers. At other times, it doesn't matter if a phenomenon is universal, only that there is evidence that it exists.

The Priority of Question Formulation

Of course, the 1 – 2 – 3 scheme is a model, and hence a simplification. It is rarely possible to specify the question without also thinking about evidence, which entails thinking about what data might be gathered, and how it might be analyzed. In practice, design of empirical studies is a highly iterative process running round the question, evidence, and data collection/analysis loop many times. Even so, there are times when a constrained opportunity demands action before adequate planning is possible: "have data set, will investigate". Sometimes one seizes the opportunity before the question comes into focus. This is a dangerous, albeit sometimes necessary, route.

From the outset, it is important to *prioritize* the framing of the question, and to use consideration of evidence and techniques as a way to *refine* the question, not as questions in their own right. The corollary is to avoid commitment to techniques until the question comes into satisfactory operational focus, and especially to beware the 'tail wagging the dog': premature commitment to a technique in the absence of a well-formed question.

Table 3: Some classic pitfalls of empirical studies. Most of these can be addressed at the planning stage. The prioritization of question formulation coupled with good, early reasoning about what evidence is required do much to avoid these traps.

---

looking before leaping
> Failure to reflect.
> Failure to recognize available evidence.
> Failure to consider conflicts, confounds, representativeness, limitations, etc.

premature experimentation
> For a precise study, you need a precise question.
> If your starting point is too complex, broad, or poorly articulated, your question will disappear 'as sand through the fingers' as you try to refine an experiment design.
> *scarcity of theory*
> Failure to explicate conceptual underpinnings.
> Failure to consider alternatives.
> Failure to contribute evidence that can accumulate or be compared.

lack of situation
> Ignorance and isolation are the enemies of discourse.
> This is not just 'bad form', it can lead to 're-inventing the wheel'.
> A day in the library can save six months of redundant research.

borrowing methods out of context
> Can lead to major oversights, and to mis-matches between method and needed evidence.
> Need to understand the underlying stance and assumptions associated with a given method. Are you applying it as it was intended? Is it able to uncover the sort of evidence you need?

putting the cart before the horse
> Choosing techniques before understanding the question.

great expectations – taking too big a bite
> 'A life's work takes a lifetime, but it is achieved one step at a time'
> 'How can one eat an elephant?' If the question is intractable; ask a smaller question.

confusing anecdote with fact
> What 'everyone knows' is not always accurate or valid.

confusing statistics with rigour
> Einstein: "Not everything that counts can be counted, and not everything that can be counted counts."
> The point is to know what can and cannot be shown with different sorts of evidence.
> The false seduction of the definitive experiment – experimentation is inappropriate when the questions are not yet precise enough.

death by surprise
> Lack of respect for failure leads to false claims, distorted reports, and loss of crucial information.
> A good study is one that is *informative*, even if it doesn't go as expected.
> Some of the most valuable results are surprises and side-effects.
> Consider in advance what 'failure' will indicate, what will happen if the study goes wrong.

4

Link Research to Relevant Theory

Interdisciplinarity and the "Trading Zone"

CS education research is inevitably interdisciplinary. The nature of CS, and of the knowledge we aim to transmit in the course of education, is rooted in mathematically-derived, computational, analytic science. However, the circumstances of the classroom, the nature of education, and models of teaching and learning, are areas that are amenable to investigation only through the human sciences. This means that our specific area is theory-scarce and we have to look to other disciplines for a theory base. The tensions of different perspectives make coherent integration of the components of research—question, theory, method—tricky. At worst, this can mean inappropriate use of "borrowed" ideas and techniques. At its best, however, CS education can resolve these tensions into a new and distinctive way of working. Peter Galison (Galison, 1997) has a construction of how new areas of working can arise:

> I intend the term "trading zone" to be taken seriously, as a social, material, and intellectual mortar binding together the disunified traditions of experimenting, theorizing, and instrument building [in subcultures of Physics]. Anthropologists are familiar with different cultures encountering

one another through trade, even when the significance of the objects traded-
and of the trade itself-may be utterly different for the two sides.

He makes the case persuasively that such trading zones have emerged (and
diminished) within Physics[5], and he identifies whole disciplinary areas that have
emerged and flourished over time.

An example of such a trade, which illustrates the separation of the parties' points
of view, is the selling of Manhattan, New York for a few dollars. The evidence for
this trade resides in a letter in the archives in the Rijksarchief in The Hague. Peter
Schaghen wrote this letter in 1626 to his employers, the (Dutch) West India
Company. In it he reports (amongst other things):

> … our people are in good spirit and live in peace. The women also have
> borne some children there. They have purchased the Island Manhattes from
> the Indians for the value of 60 guilders.

So, it would seem from the perspective of the Dutch settlers, Peter Schaghen, the
West India Company, and, perhaps especially from our historical perspective, that
this was a very good trade for the Dutch. However, the idea of land ownership did
not exist among Native Americans. The Lenape (the Native American tribe in the
area) were "trading" the right to use the land – which everyone had *as a right* – for
money and goods. The trade goods were valuable; they used uncommon raw
materials and were troublesome and time-consuming to produce. From their point of
view, the Lenape too made a very good trade. Here is a clear example of the
significance of the things traded, and the trade itself, being totally different from the
two sides[6].

However, the example of Manhattan also illustrates that trade often occurs
between unequal parties, one more powerful than the other. Ann Brown, a "classic"
psychologist who made great contributions to educational research, identifies similar
problems within a disciplinary context. She describes some of the tensions within
herself and her own research.

> As a design scientist in my field, I attempt to engineer innovative educational
> environments and simultaneously conduct experimental studies of those
> innovations. This involves orchestrating all aspects of a period of daily life in
> classrooms, a research activity for which I was not trained. My training was
> that of a classic learning theorist prepared to work with "subjects" (rats,
> children, sophomores), in strictly controlled laboratory settings. The methods
> I have employed in my previous life are not readily transported to the
> research activities I oversee currently (Brown, 1992)

She also describes how problems of the perception of the "place" in which she
chose to work was perceived by others:

> Indeed, the first grant proposal I ever had rejected was about 10 years ago,
> when anonymous reviewers accused me of abandoning my experimental
> training and conducting "Pseudo-experimental research in quasi-naturalistic

settings" This was not a flattering description of what I took to be microgenetic/observational studies of learning in the classroom (Brown, 1992)

Sometimes, of course, trading zones can result in new cultures and new ways of working. For example, the entire field of Artificial Intelligence (AI) from its very beginnings in the Second World War, has grown from the intersection of theories from philosophy, psychology and linguistics (Edwards, 1996). But, as an area of enquiry, it would not have been possible at all without "trade" between these established disciplines and computational sciences and the technology of computation (Agre, 1997). AI is fundamentally about crafting models, about building things; its medium is computers and computation. In this trading zone, the partners brought, valued and traded different things but the resulting context was more evenly balanced between their inputs.

The Trading Zone and CS Education

For a practicing scientist, even though he cannot be familiar with more than a small territory of science, must be prepared at any time to make excursions into collateral regions to find what he wants; must know, therefore, their main landmarks and enough of their languages to ask his way in them. (Holmstrom, 1947)

It may be that every interdisciplinary field is a "trading zone" (or grows from one). For CS education, we must learn to speak with our trading partners. Our use of theory is largely from the social and learning sciences. If we call upon "constructivism" would an educationalist recognize the concept as we use it? Or, if we wanted to "make an excursion" (as Holmstrom would have it) into education, would we know the terrain, would we recognize the important landmarks? Would we be able to tell what were important results from mediocre ones?

Our methods for empirical study are widely drawn—If we use quantitative methods, would a statistician recognize the "truth" of our conclusions? If we use ethnographic investigation, would an anthropologist recognize the validity of our methodology?

A different aspect of the trading zone is the necessity to play back into our own discipline. We must engage with the validity of methods in terms of the originating discipline, but also (when we engage with our CS colleagues) the validity of the investigation (and its methods and conclusions) within CS. Can we tell a "CS story" about research that we're doing?

So one sense of trading zones, of trying to build up a legitimate, well-founded, thoughtful use of interdisciplinarity, means that we have to establish standards. Part of what we need to do in order to ensure that CS education research is effective and is useful is to ensure that we understand what rigor means for us. If we understand *why* the methods that we're using are valid methods, if we understand *how* we're constructing knowledge in this way, then we can establish standards and our own distinct disciplinary norms and practices. David Schkade outlines this necessity well, in regard to his own "interdiscipline" of Information Studies (IS):

> ... an important issue is the concept of "reference disciplines". Researchers in other disciplines cannot be relied upon to develop theories that are directly relevant to IS out of the goodness of their hearts. Their objectives are different. The IS field must develop at least some researchers who are competent theory builders in their own right, so that existing theories in reference disciplines can be rigorously adapted, or new theories developed, as needed. There is a lot of bad psychology, economics, etc. out there as well as good and useful work. (Schkade, 1989)

When human cultures engage in trade, they often do not speak a common language, but a derivative of the language of the more dominant partner, which has a much reduced vocabulary and simplified grammar. Such trading languages are called "pidgins". A pidgin language, when it develops native speakers—when children are born who use it as their first language—grows in sophistication and complexity, developing new vocabulary, structures and idioms, and is then called a Creole. CS education research is, today, a pidgin. The outstanding question is whether we can develop distinctive areas of working—whether we can develop toward a Creole. We conclude with David Schkade: "Research imported from other disciplines should be viewed with a healthy scepticism until ... researchers, developing the reasoning on their own, see the rationale and applicability themselves."

Theories of Learning: A Reference Discipline and a Trading Partner

If we are to take seriously the guiding principle "link research to relevant theory", and if CS education is theory-scarce, then we must be familiar with the construction and investigation of theory in other, relevant, areas. One of the most obvious territories to explore is that of learning theories. These have themselves been formulated in several separate disciplines: from cognitive and educational psychology, certainly, but also from sociology and social policy and from developmental and childhood studies. Also, historically, there have been many influential individuals who have founded schools—and schools of thought—and accrued followers. We need to be aware of their histories and traditions before we claim them as trade goods.

There are different ways in which researchers can think about education and learning, and different impacts in which they might impact on research agendas. Some of these ways come directly from theories/theorists. Others come from the ways in which theories have been instantiated in educational environments. This separation is crude but indicative. We explore further a few possibilities of the ways learning theories have been used.

Learning

Empirical laws

There is considerable work within "classic" cognitive and developmental psychology that examines and defines broad categories of human cognitive capacity. These investigations and results have a significant effect on human learning, even if there is little we can do to address them with any specific educational (instructional) context. For example, the finding that human short-term (or "working") memory is more-or-less bound to the limit of seven (plus or minus two) things (Miller, 1956) is obviously of interest in an educational setting. Another example is the acquisition of implicit learning, which requires large numbers of instances with rapid feedback about which category the instance fits into (Seger, 1994); wittingly or unwittingly, we can produce an environment, especially in software, that affects how students learn. In an analogous way, the work of Mihaly Csikszentmihalyi on the notion of "flow", the necessary conditions for human beings to be fully engaged in an activity (Csikszentmihalyi, 1991), informs the milieu of electronic game design, without actually providing a one-to-one relationship with specific elements.

Theories

There are many educational theories, from a variety of sources. For example, the work of Jerome Bruner (which builds upon the structured stages of cognitive development outlined by Jean Piaget) emphasizes the relationship of cognitive structure to the structure of disciplinary content: "What are the implications of emphasizing he structure of a subject, be it mathematics or history—emphasizing it in a way that seeks to give a student as quickly as possible a sense of the fundamental ideas of a discipline?" (Bruner, 1960).

Alongside Bruner's (and often associated with it) is the work of Lev Vygotsky. Vygotsky's ideas centered on the notion that knowledge and learning are culturally and societally constructed. In particular, his idea of the "zone of proximal development" (ZPD) has been enormously influential. ZPD states that students have limitations in the amount of progress they can make from their current knowledge state, but, with the help of a teacher giving appropriate interventions and scaffolding, their understanding can expand further than it would if they were left alone. "…the distance between the actual level of development as determined by independent problem solving [without guided instruction] and the level of potential development as determined by problem solving under adult guidance or in collaboration with more capable peers". (Vygotsky, 1962)

The general ideas represented by these (and other) theorists taken together lead to an understanding of what happens in teaching and learning that has been labelled "constructivist": encapsulating the idea that learners actively construct knowledge, rather than stand as passive recipients, as vessels to be filled. This can lead directly to classroom implications, entailing specific practices such as "reciprocal teaching" (Brown & Palincsar, 1989; Palincsar & Brown, 1984) and "jigsaw instruction" whereby students learn through constructing their knowledge in order that they can teach others; or to more general approaches, for example as outlined by Moti Ben-Ari (Ben-Ari, 2001).

Another collection of approaches, known broadly as "behaviorist", grow from the work of B.F. Skinner (Skinner, 1938) and focus only on objectively observable behaviors, therefore discounting internal mental activities. In a behaviorist environment, learning is considered to be the acquisition of new behavior. "Conditioning" for teacher-approval, high marks, or other reward, is the process by which learning occurs. Although they find some use in the classroom, these ideas are more often seen incorporated into various on-line pedagogic environments (Skinner, 1968).

For CS education, the work of Seymour Papert, co-founder of the MIT Media Laboratory and collaborator of Jean Piaget, is influential (Papert, 2003). His theoretical approach holds that things that are readily available in the everyday environment provide relevance and concrete experience on which learning is constructed. In order to develop mathematical reasoning, therefore, it is important to provide relevant stimuli in the environment, together with language for discussion of the resultant concepts. The tradition of instructional approach which has developed from this is focused around the use of LEGO in the classroom, although there are other "constructionist" approaches which do not rely on proprietary manipulables.

More recently, the work of Jean Lave and Etienne Wenger (Lave & Wenger, 1991; Wenger, 1998) has extended the notion of the social nature of learning with ideas that learning is always situated within authentic situations, and takes place within communities of practice. With this is associated larger notions of how communities are structured and how learning occurs in them, how learning and membership of community are closely identified with each other and how knowledge cannot be separated from practice. This is an intriguing theory for CS, a discipline that has a clear set of vocational communities and constituencies, and it is one which has begun to be thoughtfully explored within the CS classroom (Kolikant, in press).

Models and taxonomies

As well as broadly-conceived theories which influence at the most general level, there are also more narrowly-drawn concepts, more precisely targeted at areas or types of education.

Bloom (et al.)'s Taxonomy, first codified in *Taxonomy of educational objectives, handbook 1: The cognitive domain,* describes a range of cognitive behaviors found in educational assessment. The taxonomy comprises six levels, arranged along a continuum of complexity, *vis*: knowledge, comprehension, application, analysis, synthesis, evaluation. It was devised in order to describe educational objectives that went beyond mere recall of fact and was aimed at "teachers, administrators, professional specialists and research workers" and was "especially intended to help them discuss these problems [of educational objectives] with greater precision" (Bloom, 1956). It has been abidingly influential, and has recently experienced a renaissance in CS education research (Lister & Leaney, 2003).

The use of Kolb's Learning Cycle (Kolb, 1984), on the other hand, is to structure instruction so that experience is seen as the source of learning and development.

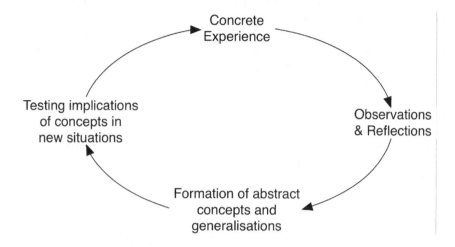

Figure 3: The four stages of Kolb's Learning Cycle

In contrast to these cognitive models, William Perry relates learning to a model of intellectual development maturity (Perry, 1981; Perry & Harvard University. Bureau of Study Counsel, 1970). Perry claims that college students (and others, too) "journey" through nine "positions" with respect to intellectual and moral development. These stages can be characterized in of terms the student's attitude toward knowledge. There are nine levels, grouped into four stages: dualism, multiplicity, relativism, commitment[7].

Table 4: Perry's Model of Intellectual and Ethical Development

Stage	Characterization	Students' View
A. Dualism/Received Knowledge:	There are right/wrong answers, engraved on Golden Tablets in the sky, known to Authorities	The students' task is to learn the Right Solutions learn the Right Solutions and ignore the others
B. Multiplicity/Subjective Knowledge:	There are conflicting answers; therefore, must trust one's "inner voice", not external Authority	The students' task to learn how to find the Right Solutions
C. Relativism/Procedural Knowledge:	There are disciplinary reasoning methods	The students' task is to learn to evaluate solutions
D. Commitment Constructed Knowledge:	Integration of knowledge learned from others with personal experience and reflection	The student explores issues of responsibility The student realizes commitment is an ongoing, unfolding, evolving activity

The journey is sometimes repeated, and one can be at different stages at the same time with respect to different subjects.

The aspect of learning that they have chosen, what they have chosen to emphasize in the creation of their model, what to simplify, and what to discard—all contribute to the differences between these constructs, but Bloom, Kolb and Perry have all devised *models*.

Models have also been developed with regard to situated classroom instruction. One of the most enduring in recent times has been the model of *Problem Based Learning*, first instantiated at McMaster University Medical School as a method of helping students learn skills of medical diagnosis (which were poorly served by lectures and other, formal, classes). Problem-based learning has spread widely in CS education.

Instruments

Finally, at the lowest level of granularity, there are specific instruments devised to expose particular aspects of the instructional situation. Mostly these are manifest as questionnaires or scales which describe students. For example there are inventories on learning styles (Kolb, 1984), on approaches to studying (Entwistle & Tait, 1995) and on personality types (I. B. Myers, 1998, 2000). Their use within the classroom is predicated by the instructor being interested in a very specific idea, for example, in adjusting materials to groups of student with differing learning styles, or for assuring well-formed groups for project work. The step that follows—of evaluating whether there is a difference between the groups distinguished by the instruments—is the step toward their use in CS education research.

Summary

We believe that the way in which CS education *links research to relevant theory* is via trading zones and reference disciplines. It would be impossible to be comprehensive about sources of theory; indeed, it is difficult to be comprehensive about material within a single source. Our limited survey above indicates some of the ways in which CS education as a research area overlaps with, draws upon and "trades" with education.

The theories, methods, instruments and measures located within our reference disciplines have different traditions and uses. How they are represented and used within our disciplinary domain, a necessarily interdisciplinary context, is a source of both problem and opportunity for CS education research

5

Provide a Coherent and Explicit Chain of Reasoning

It is not enough that we feel confident in our work; we must be able to explain it to others—in the classroom probably, in the laboratory possibly, in print certainly. "Science" is about sharing, and if the people you want to share your material with cannot follow the progression of your argument, cannot understand the reason that you chose the question in the first place … cannot understand the reason you think your choice of method is going to provide compelling evidence (even *sufficient* evidence) … cannot understand the relationship you claim for your intervention and the cited theoretical tradition … then this not just bad form, it is bad science.

So we need to "provide a coherent and explicit chain of reasoning". *Coherent* and *explicit* are straightforward and understandable terms.

Our accounts must be coherent because incoherent accounts are difficult to follow and inherently suspect.

Our accounts must be explicit for three reasons. Firstly, because we seek to be rigorous and precise. If we do not describe—precisely—what we did, and do not provide—precisely—the background information, and do not detail the assumptions we built on, then our work cannot be judged fairly. It may not be bad work, but it will be impossible to tell. Our accounts must be explicit, too, because we aim to make them comprehensive. What is un-stated will be unclear, and so will be susceptible to many different interpretations. Finally, we have a particular duty to be explicit in our work as, in an immature discipline, we do not have a set of common, taken-for-granted knowledge or way of working.

However, *chain of reasoning* has different implications when used in different contexts, and if we ask what a chain of reasoning *is*, the reality of working in a trading zone may cause problems. This is graphically illustrated by Marian Masterman's comment on Thomas Kuhn's *The Structure of Scientific Revolutions*:

> Insofar, therefore, as his material is recognizable and familiar to actual scientists, they find his thinking about it easy to understand. Insofar as this same material is strange and unfamiliar to philosophers of science, they find any thinking that is based on it opaque. (Masterman, 1970)

Whatever view you take on Kuhn's book, the chain of reasoning is the same in both cases; yet it demonstrably causes problems for an unfamiliar audience.

However, if we turn our heads slightly and ask what a chain of reasoning *does*, the meaning becomes clearer. A *chain of reasoning* is the strong intellectual filament upon which we can string the pearls of our work. In that sense, it becomes possible to determine some fairly standard constituent components.

Relationship to theory

Theory can, in broad terms, be used in two ways. We can situate our work in exploration of a theory-base (or theoretical position), or we can use a theoretical perspective to inform our investigations. There is a difference between seeing the world/classroom from a certain perspective and designing the world/classroom according to particular perspective. For example, if our work were an exploration of theory, we could imagine asking questions (and seeking associated evidence) along the lines of "Are CS educators behaviorists?" (or constructivists, or whatever), or perhaps "Are their practices behaviorist—with or without their intention?" If, however, we were conducting work that was informed by these same theories, then we would be using them as a conceptual lens and, in looking from a behaviorist perspective, we might expect to illuminate certain qualities in any teachers' practice.

The first step in our chain must be to articulate the relationship that our work has to any theory, or theoretical assumptions, and how our work uses or interprets those.

The "chain" in chain of reasoning

In talking about research studies, there are several components, at different levels of granularity, that have to be considered. (See Figure 4.)

The chain starts, as explored above, with theory. However, theories don't stand alone and are related to disciplines (as well as each other). Theory and discipline taken together can inform the research questions we ask. Often, situation within a disciplinary context brings with it an associated methodology, which prescribes the selection, combination, and sequencing of the methods and techniques we employ.

So, asking questions within an economic context will force the use of a different methodology than asking them within a psychological or sociological context.

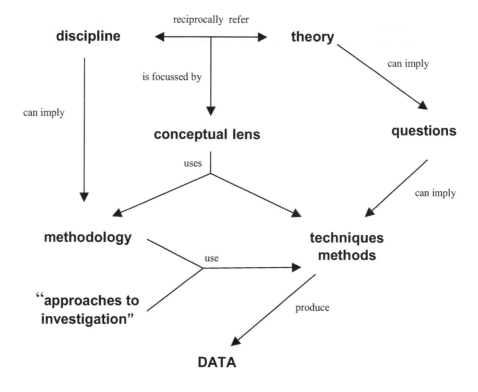

Figure 4: Some components (the labels) and influences (the labelled arrows) to consider in a chain of reasoning.

A methodology is not the same as any specific method or technique. For example, "diary studies" or "questionnaires", both useful methods, may be included in several distinctly different methodologies.

There are also some specific aggregations of methods that might be termed "approaches to investigation". Phenomenography (Research, 1995), activity theory (Engestrom, Miettienen, & Punamaki, 1999; Nardi, 1996) and grounded theory (Glaser & Strauss, 1967) are all examples of such approaches. These are not tied to a particular discipline. They are characterized by a stance which focuses the nature of inquiry (e.g., phenomenographic investigation seeks patterns across individuals via in-depth qualitative techniques), a selection of methods oriented to the stance, and often a descriptive language or framework. They are "approaches to investigation" because they are a package; they cannot be used in a "pick and mix" way, taking just one part of them. Use of such an approach constrains both data gathering and analysis.

So, what the "chain" in "chain of reasoning" *does* is to detail and describe each of the possible points of situation, implication and dependency where choices, or assumption, have been made. The end result of all these implications, choices and dependencies is data. A strong chain of reasoning conveys many benefits onto our resultant data: assurances of validity, replicability and representation for example, as well as higher confidence that the study has not been biased.

> Even the most rigorous empirical methodology is no substitute for careful development of the reasoning that underlies the hypothesis (Schkade, 1989)

6

Use Methods that Permit Direct Investigation of the Question

The first key principle is *pose significant questions that can be answered empirically*. In our pragmatic approach, the question and evidence requirement taken together inform the choice of research method or technique (the terms are, for pragmatic purposes, interchangeable). The technique must deliver data of a sort that can answer the evidence requirement—that is fit for the purpose. This means that the technique must generate data (and hence evidence) that is *sufficiently rich*, that it must provide enough information to address the research question *at the right resolution*, and that it must be *feasible* within available resources.

Richness of Data

One of the factors that characterizes an empirical technique is the richness of the data it is likely to produce. *Richness* here is used to mean the number and generality of questions that the researcher can attempt to answer using that data. For example, examination scores offer single-point data that can only be used to answer limited questions, whereas the examination questions and student answers taken together would offer much richer data, and the questions and answers taken in conjunction with interviews with the students about why they answered as they did, would be richer still.

Richness of data potentially translates to richness of evidence, depending on the method of analysis and on the interpretation based on those analyses—on inferences made from the analyzed data. Alternatively, the data collection technique and method of analysis may reduce the richness of the data, by aggregation or selection. For example, an interview, a method which yields inherently rich data, can be combined with a coding scheme which records selected data, such as categories of activity against time. Rich data may also be constrained by aggregation, for example averaging across subjects or sub-populations.

Level of Resolution

The nature and focus of the research question implies an equivalent level of detail and specificity in the data. *Fitness for purpose* requires a match between granularity of the research question and the level of resolution of the data collection, and hence requires a method that can generate data of appropriate resolution. For example, studies of individual behavior give insufficient insight into social interactions, and surveys of group process give little insight into individual cognition. Mantei (Mantei, 1989) distinguishes five levels of resolution of question and data, as summarized in the *Levels of Resolution* table.

Table 5: Levels of resolution

Level	Research question focus	Resolution of data
micro-micro	questions about internal structures and processes of the human mind (e.g., memory, cognitive load)	performance measures reflecting internal cognitive mechanisms, (e.g., response times in microseconds)
micro	questions that focus on the individual's interaction with the external environment (e.g., how individuals use tools to solve problems)	specific data about individual behavior, such as sequences of decision making and problem solving (e.g., , how individuals perform given tasks)
Standard	questions about regularities associating individual characteristics with individual behavior (e.g., effects of personality differences on productivity)	micro-level data, as well as data about attitude, style, and preference, aggregated for a given individual at the time of data collection, individual data aggregated over a group of individuals (e.g., averages of individuals' performance)
Macro	questions about group properties, behaviors, and processes (e.g., group creativity, leadership, cohesiveness)	group-generated data on group behavior
macro-macro	questions about systems, networks, and organizational behavior (e.g., impact of computer-supported meetings on organizational communication patterns)	data aggregated over a group of people who do not interact with each other during the data collection; aggregation of responses from multiple subgroups and individuals

Analysis methods must match the data resolution in order to provide meaningful evidence at a level appropriate to the research question. A mismatch between analysis and question can produce findings of dubious utility; the findings may have some meaning, but they are unlikely to address the question effectively. Mantei [ref] gives an example: "Cognitive style measures in IS have been criticized because they serve as the wrong level of data collection for answering questions about the design of an information system's interface, which affects behavior at the micro level."

Costs

A number of factors characterize a technique. Among the most important are:

- location (in situ, in lab)
- what data (and richness of data)
- how much data
- level of resolution of data collected
- number of subjects
- representativeness of subjects (which has implications for generalizability of results)
- whether the research strategy is theory-driven or data-driven
- basis of analysis: descriptive, statistical, etc.
- precision of question (maturity of investigation)
- constraint of conditions / number of variables

Unfortunately, data doesn't come "for free". Any data implies data collection and analysis costs. The more and richer the data, the higher the cost. Costs arise at all stages of an empirical study: from design and planning, through implementation and execution, to analysis. Control, validation of instruments, and pilot studies all contribute to planning cost, and all influence the quality of the data collected. The preparation of study materials, gaining access to subjects and settings, and conducting the study, all contribute to implementation and execution costs. All resources incur costs: time, subjects, settings, researchers, instruments, equipment, expertise, etc. Cost implies constraint: few researchers have unlimited resources, and hence budgets constrain research designs.

Tradeoffs

Richness and resolution trade off with cost. Mason summarizes: "Ultimately, the total resources available to conduct an experiment delimit the amount of knowledge that can be obtained from it." (Mason, 1989). Hence, within a given level of resource, individual factors in the study design trade off against each other. For example, the cost of data collection and analysis in a case study may trade off against the number of cases that can be considered. Increasing the precision of the research question can afford an increase in sample size.

A balance of tradeoffs in study design is assessed in terms of fitness for purpose. If the evidence is required to be a statistically significant correlation, then the sample size must be sufficiently high, and the number of variables under consideration may be constrained. If the evidence is required to be a contextualized account of a problem-solving process, then data collection and analysis costs may be unavoidable, but fewer subjects may be sufficient. The aim must be to amass evidence of sufficient utility within the constraints of cost—and the aspiration must be to amass evidence of the highest possible utility within the budget

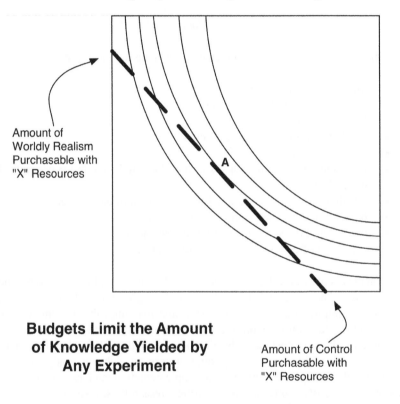

Budgets Limit the Amount of Knowledge Yielded by Any Experiment

Amount of Worldly Realism Purchasable with "X" Resources

Amount of Control Purchasable with "X" Resources

A

Figure 5: Mason's (Mason, 1989) depiction of how studies are constrained by costs. Mason uses iso-episteme curves to illustrate how realism (y axis) trades off with control (x axis) and suggests that budgets (the dotted line) limit the amount of realism or control that can be achieved. He suggests that studies that lie at point "A" on the curve will be most cost-efficient. However, "fitness for purpose" may demand a different strategy.

Settings and Configurations

Techniques may be used in a variety of settings and in a variety of configurations:

Settings
Techniques can be applied *in situ* (in the normal or 'natural' environment), for example, studying professional programmers in their workplace, or studying students in the classroom; *under constraints* (in a natural environment on which selected limitations have been imposed) for example, studying students in their familiar classroom setting using a particular programming environment; or *in a laboratory* (a highly-controlled environment).

Configurations
Methods are not only used in a "one-off" manner, but can be employed (and re-employed) in various configurations:

* A single technique may be applied in different ways to the same research question, refined through successive iterations. The refinement may be, for example, by way of sample size, or choice of task, or method of data recording.
* A single technique may be applied at many levels: the scope of research can vary from investigation of a tool, to a course, or a whole curriculum.
* A single technique may be employed in longitudinal fashion; for example, concept acquisition might be studied at intervals within an entry-level programming class. The cohort might also be re-visited a year later, perhaps after a comparative languages course.

Methods / Techniques

The purpose of this section is simply to indicate the range of research methods or techniques available, not to provide a complete catalogue nor to endorse any particular approach. A discipline-based methodology might specify techniques. In CS education, in the absence of such a methodology, the choice of technique is influenced by the research question. Whether the research is theory-driven (testing hypotheses derived from theory which predicts the outcome) or inductive (seeking emergent patterns in the data) can shape the techniques, but the techniques themselves do not pre-suppose a given approach or methodology.

Case studies
Case studies are in-depth, descriptive examinations, usually of a small number of cases or examples. They provide an intensive, holistic description of a single phenomenon, investigated in situ. Case studies usually encompass a variety of data collection techniques, potentially ranging from ethnographic and participant observer methods, through artifact analysis, through interviews, to constrained tasks. Case studies are appropriate especially when the boundaries between the phenomenon and the context are not clear, when the objective is to tease out as many factors contributing to some phenomenon as possible. Because of the numbers and sensitivity to context, there are limitations to generalization of findings. Analysis tends to be inductive reasoning, based on multiple data sources. Case studies are demanding and intensive.

Table 6: Case study tradeoffs

Good for:	Case studies are appropriate especially when the boundaries between the phenomenon and the context are not clear, when the objective is to tease out as many factors contributing to some phenomenon as possible.
Bad for:	Generalization, given small numbers and sensitivity to context.
Kind of evidence:	Very rich and contextualized. Group or individual. Possibly historical. Small numbers.
Cost of planning:	Can be low, depending on how focused the study is.
Cost of data collection:	Tends to be high, because it usually involves in-depth interviews and observation, although some types of data (such as existing records and artifacts) can be low-cost.
Cost of analysis:	High. A good case study requires considered analysis, and integration of evidence from multiple data sources. There is often no pre-specified protocol.

Diary studies

Diary studies rely wholly on self-report; individuals are asked to keep recorded accounts of their behavior over time. Diaries may be structured, to focus reports on key issues and to facilitate comparisons. Diary studies afford a glimpse into subjects' introspection over time. Introspective accounts can provide considerable insight, but they suffer a number of limitations. For example: individuals vary enormously in their value as diarists; reflecting on behavior can influence subsequent behavior, changing the very phenomena under observation; diaries are selective, usually retrospective and rationalized. Hence, diary studies are usually combined with other data sources.

Table 7: Diary study tradeoffs

Good for:	Insights into individual experience, perceptions, and beliefs. Afford good potential for longitudinal views.
Bad for:	They present a high chance of distortion, being dependent on individual skill, and often selective, retrospective and rationalized accounts. Can be intrusive on natural behavior.
Kind of evidence:	Rich, individual, often longitudinal accounts. Numbers usually small.
Cost of planning:	Low to medium. Planning cost depends on whether the diary is structured.
Cost of data collection:	Low. Collection is via self-report, so the cost derives mainly from whatever incentives or spurs are provided to keep the diarists active.
Cost of analysis:	Very high, as with any analysis of rich qualitative data.

Constrained tasks, quasi-experiments, and field experiments

Constrained tasks are intended to bridge between observation and laboratory experiment, providing some constraints on subjects' activities (and hence some basis for comparison), while maintaining the richness of context. The level of constraint varies; typically, specified tasks are carried out in situ (hence the term 'field experiments'). The constraint is usually on the task, which is chosen to represent some aspect of natural activity, in order to investigate some phenomenon of interest—usually one identified in previous observation. Sometimes the constraint is on the environment, in order to tease out factors contributing to behaviors or processes of interest. The constraint may be stronger, as well, so that the study draws on experimental techniques, and seeks quantitative data, but without full experimental control (hence the term 'quasi-experiments'). For example: a classroom-based study styled on experimental comparison but undertaken in situ. One of the limitations of quasi-experiments is that numbers may be insufficient to exclude environmental factors through statistical analysis. Another variant is to try to reproduce a typical environment in a laboratory, for example providing a representative designer's library and typical range of tools and media, and specifying a design task. Constrained tasks offer some increased control over observation and hence provide some basis for comparisons and for validation of observations—while maintaining some realism. However, they do not have the power or precision of laboratory experiments, and generalization is limited.

Table 8: Constrained task tradeoffs

Good for:	Bridging between observation and experiment, for example in order to investigate an observed phenomenon in more depth and with more control, without stripping away context. Helpful in focusing in on key factors and their inter-relationships. Can provide a basis for comparison among subjects.
Bad for:	Generalization is limited, due to limited control and preservation of context. Often numbers are small.
Kind of evidence:	Regularities associated with the particular task, whose identification may be supported with statistical analysis.
Cost of planning:	Medium to high, approaching the planning cost associated with controlled experimentation. Care is required for the selection of the task and subjects, the constraint of the environment, and the specification of the protocol.
Cost of data collection:	Variable, depending on what data is collected. Constrained tasks may use quantitative measures (which have a relatively low collection cost) or rely on qualitative data (which can entail higher collection cost).
Cost of analysis:	Variable. Can be low for quantitative measures, high for inductive analysis of rich qualitative data, or medium for a focused, mixed analysis.

Document studies

Existing records, logs of electronic communication, individual notes and diaries, sketches and diagrams—various written or recorded artifacts provide a naturally-occurring source of information, often closely allied with a phenomenon of interest, and potentially offering insights into processes, interactions, organizational character and culture, and individual experience that may be hard to capture otherwise. They are steeped in the context in which they are produced, reflecting both environment and language, and they can provide a longitudinal view. They can provide an unobtrusive form of data collection. The utility of such "documents" depends on their completeness, authenticity, accuracy, and representativeness. Documents may well be colored by the purpose for which they were originally produced; this can be an advantage, or a limitation, depending on the focus of the study. The analysis of a "corpus" or collection of documents is time-consuming and demanding, and it may be difficult to assemble an appropriate, representative corpus.

Table 9: Document study tradeoffs

Good for:	Unobtrusive, longitudinal, context-sensitive views of phenomena of interest.
Bad for:	Limited by what documents are available, with implications for completeness, accuracy, and representativeness. There is often no access to interpretation by the originators of the documents, and hence gaps may be hard to fill.
Kind of evidence:	Very rich, contextually steeped, qualitative material often affording a longitudinal view and multiple perspectives.
Cost of planning:	The planning cost is associated with planning the analysis and is dependent on the nature and variety of documents under consideration.
Cost of data collection:	Can be very low, because corpora may be available for 'harvest'. However, the challenge lies in acquiring an appropriate, representative corpus for the purpose.
Cost of analysis:	Very high, although it can be ameliorated by pertinent automated tools.

Automated logging

Any computer-mediated activities can be recorded automatically, for example as activity logs, streams of keystrokes, or sequences of electronic communication. Such logs can be comprehensive (for the data they collect), precise, and accurate. They can include precise timing information. Electronic logs in appropriate formats are readily amenable to automated analysis of many varieties, from performance measures to linguistic profiles. From automated logs, it can be possible to reconstruct with detail and accuracy the conduct of a task by many subjects, for example to analyze sequences of actions, associations between actions and errors or actions and outcomes, and time spent on different task components. Automated logs facilitate performance comparisons between subjects or activities in terms of speed, accuracy, and outcomes. Their disadvantage is that, although they record precisely what people do while they interact with the system, they offer no direct information about what they intended, or where they looked, or what they did when they weren't interacting with the system. Although they can record the electronic context well, they record only the electronic context, potentially omitting factors important in the phenomena of interest. However, automated logging combines well with interview techniques which give insight into intentions and personal experience.

Table 10: Automated logging study tradeoffs

Good for:	Unobtrusive, accurate capture of electronic communication and interactive behaviors, including precise timing of actions.
Bad for:	Relating behavior to intention.
Kind of evidence:	Can be qualitative or quantitative accounts of behaviour, communication and interaction, usually supported with statistical analysis.
Cost of planning:	High. Logging easily produces a flood of precisely detailed data; planning the collection and analysis of that stream requires careful reasoning about how to interpret research questions and how to filter and manipulate that data relevantly.
Cost of data collection:	The collection cost is associated with the creation of logging tools, and subsequently with the cost of data storage. Once the tools are in place, the collection cost is minimal.
Cost of analysis:	Can be low, depending on the question and the available tools. Can be high, depending on the level of human intervention and interpretation, and on the need for new or customized tools.

Observation

Observation is an extremely broad category of investigation, ranging from intensive ethnographic methods through targeted, short-term approaches. The common theme is the watching—and recording—of behavior in context, usually in a natural situation and environment. Observation can produce data that is descriptive (e.g., a record of behavior, possibly within a descriptive scheme), inferential (considering the intentions behind observed behavior), or evaluative (assessing or measuring characteristics of observed behavior). Hence, two key aspects that distinguish different observation approaches are the level of participation of the observer, and the nature of the records kept (especially whether the records preserve the richness of the setting or focus on selected phenomena).

Records can take many forms, from ethnographic field notes, through verbatim contemporaneous notes, to audio and video recording. The data can be descriptive or quantitative. The obligation is for records to be complete and accurate. Although some records (such as field notes) might be made immediately after-the-fact, they must ultimately stand on their own, and hence must provide a sufficient record without reliance on additions from memory.

Observation can produce very rich, highly-situated data reflecting behavior in context. It can provide opportunities to identify important factors which were previously un-remarked. It can capture complex interactions in a rich social, physical, and activity environment. However, it is demanding both in terms of data collection and in terms of analysis. Selection by the observer (or by observation protocol or the recording scheme) or expectations the observer brings to the setting may color the data, and the mere presence of the observer may influence the behaviors observed.

What follows is an indication of the variety of observation strategies that might be adopted.

Table 11: Some observation strategies

Participant observation:
> The observer participates in the respondent's natural activities (e.g., becoming a member of a design team) for first-hand experience, in order to become integrated in the social interaction and immersed in the culture. In effect, the observer becomes a collaborator or an apprentice of the informant. Insight may arise from shared or common activities. The impact of the participation may 'cut both ways': on one hand, it may distort the activity and interaction; on the other, it may reduce bias by making the interaction with the informant more naturally a part of the task.

Ethnographic observation:
> The aim is to understand the activities within the informant's frame of reference. Questions concern the correct identification of behavior. Typically, the observer comes prepared with a theoretic framework for describing what happens (for example, concepts of kinship and ritual).

Unobtrusive observation:
> The aim is to observe (and possibly question) with as little impact as possible on the informant's activity. The observations gathered tend to be descriptive, unless observation is paired subsequently with interview.

Structured observation:
> The 'unobtrusive' observer codes the informant's behavior in terms of pre-defined categories or scales.

Systematic observation:
> In this quantitative approach, observations are captured in terms of existing schema, for example, behavior might be coded behavior in terms of a set of categories, or rated on a scale.

Observation of constrained tasks:
> The aim is usually to control what the informant does, and possibly constrain the environment in which it is done, and thereby set a task which will expose some interesting or obscure part of the informant's behavior, or provide a basis for comparison between informants.

Observation of tasks with concurrent verbalization:
> The informant, having been instructed in 'thinking-aloud' or articulating normally silent processes, is asked to verbalize while performing some task.

Observation of working in pairs:
> One technique for drawing verbalizations (including explanations or articulations of reasoning or other internal processes) from the informant is to ask informants to work in pairs, so that communication about the performance of the task is inherent in the task. One assumption is that the two informants will share the same frame of reference.

Fundamentally, the quality of observation depends on the quality of the observer. Observation requires skill: in attending, in filtering, and in recording. Sometimes it also requires domain knowledge, in order to comprehend what is being observed. Whether observation is open and descriptive, or guided by a theoretical framework or by an observation protocol (a script which identifies which information is to be gathered and what criteria are to be applied), the observer needs training and experience in order to gather data consistently and accurately.

Table 12: Observation tradeoffs

Good for:	In-depth views of real phenomena as it occurs naturally. It can provide opportunities to identify important factors which were previously un-remarked. It can capture complex interactions in a rich social, physical, and activity environment.
Bad for:	Selection by the observer (or by observation protocol or the recording scheme) or expectations the observer brings to the setting may color the data, and the mere presence of the observer may influence the behaviors observed. Limited generalization.
Kind of evidence:	Rich data reflecting behavior in context.
Cost of planning:	Low. Planning cost is associated mainly with the preparation of any observation protocol, if one is used, and gaining access to the subjects in their environments.
Cost of data collection:	Very high. Data is collected only while the researcher is present; there are no shortcuts.
Cost of analysis:	Very high, although some approaches employing coding schemes or ratings reduce analysis costs significantly, at the expense of richness.

Interview

Interviews are guided dialogues, valuable in eliciting subjects' experiences, perceptions, opinions, attitudes, intentions, and beliefs. They allow subjects to respond in their own words, to explain behaviors in terms of their own values, goals, and expectations, to assign their own meanings, and to provide clarification. Interviews can elicit affective responses as well as cognitive processes. They can range from open-ended, in-depth probing of key topics to highly structured "oral questionnaires" which emphasize uniformity of the interview "script" for all subjects. Interviews are normally conducted face-to-face, but telephone and even electronic interviews can provide useful data.

The strength and weakness of interviews resides in the interaction between the interviewer and the respondent. The interview can be influenced by the quality of the rapport between the two, by the compatibility of their frames of reference, and by the skill and knowledge of the interviewer. The potential for bias or distortion is high. The questions themselves may influence responses, depending on phrasing, on individual interpretation, and on associations they may trigger. Subjects may try to please the interviewer, or to anticipate the "correct" or desired response. The quality of an interview is influenced by the subject's ability as a self-reporter: on recall, selection, and accuracy.

Yet interviews also have high potential to provide insight into people's thinking and feeling. They can be combined with other techniques in order to compare what a respondent reports in interview to what the respondent does in practice, and hence to corroborate the accuracy of reports and provide insight into behavior, motivation, and perception.

What follows is an indication of the variety of interview strategies that might be adopted.

Table 13: Some interview strategies

Ethno-methodological interviews:
> Interviews which take an ethnographic stance: the interviewer is "as a Martian", arriving (notionally) without preconceptions and seeking to elicit the respondent's meanings. There is no assumption of a shared frame of reference; the point is to elicit the respondent's frame of reference. The interviewer is non-directive; the interview is largely directed by the respondent, who maps out the topic. Probes are used to verify the interviewer's understanding.

Ethnographic interviews:
> Ethnographic interviews are concerned with eliciting the respondent's frame of reference. The interviews are open; the respondent maps out the topic, and probes verify interviewer's understanding. However, the interviewer brings to bear theories of society and interaction, and may therefore structure understanding in terms of the framework provided by the theory.

Semi-structured interviews:
> The overall structure of semi-structured interviews is planned by the interviewer in advance, with a script of main questions. The order of questions may be altered to adapt to the subject's responses; the respondent is given considerable freedom of expression, but the interviewer controls the interview to ensure coverage. Prompts (open questions encouraging breadth) and probes (focussed questions which seek to clarify or specify, to explore depth) fill in the structure.

Structured interviews:
> Structured interviews are organized according to a fixed script of carefully-phrased questions. The order of questions is fixed, and follow-up questions are minimized. The script ensures coverage and comparability across multiple interviews with different respondents.

Oral questionnaires:
> Oral questionnaires are formal, highly-structured interviews, largely comprised of closed or focussed questions presented in a fixed order. There is no additional or follow-up questioning, no deviation from the question script.

Group interviews or group elicitations:
> Group interviews add social context to the interview, allowing group dynamics to play a role in eliciting data through interactions within the group. Such interviews are usually semi-structured, with open discussion questions or group tasks. Group dynamics cut both ways: they can draw out differing perspectives and challenge individual thinking, but they can also exert peer pressure that inhibits or distorts individual response.

Focus groups:
> Focus groups target specific sub-groups, examining their responses to products, processes, arguments, etc. They are typically used in market research, where the 'focus' is on different kinds of customers. There is typically also a 'focus' on particular topics or objectives. Focus groups involve semi-structured group interviews, and they use group interaction explicitly to generate data. Participants

make individual responses, but they hear and can react to others' responses as well. The interviewer acts as a moderator who keeps the discussion focussed and ensures that all voices are heard.

Table 14: Interview tradeoffs

Good for:	Eliciting subjects' experiences, perceptions, opinions, attitudes, intentions, and beliefs. Permit in-depth probing and elicitation of detail. Powerful when it is important to understand the interaction between attitudes and behaviors.
Bad for:	The interview can be influenced by the skill and knowledge of the interviewer, as well as by the recall, perception, and self-reporting ability of the respondent. The potential for bias or distortion is high.
Kind of evidence:	Rich data reflecting what people think and feel.
Cost of planning:	Low to medium. Planning cost is associated mainly with the interview script and the analysis.
Cost of data collection:	Collection costs increase with the number of interviews. Skilled interviewers are required.
Cost of analysis:	Very high for open interviews, given the volume of qualitative information. Can be low for highly structured interviews, which limit the richness of the data.

Survey research and questionnaires

Survey research involves gathering information for scientific purposes from a sample of a population using standardized instruments or protocols. Ultimately, the purpose of survey research is to generalize from the sample to the population about some substantive issue (Kraemer, 1991).

Kraemer identifies three characteristics of survey research:
- It is a quantitative method requiring standardized information designed to produce quantitative descriptions of some aspects of a study population.
- The principal means of collecting data is by asking structured, pre-defined questions.
- Data is collected from or about a sample of the study population, but is collected in such a way as to support generalization to the whole population.

Questionnaires (or surveys), then, are a method of data collection within "survey research", as are structured interviews. They have the potential to generate large volumes of data at relatively low collection cost. Surveys can be descriptive (fact-finding, enumerating, characterizing a population) or analytic (seeking associations or causal relationships among variables).

Questionnaires typically rely on self-report: subjects' own responses to questions about their own behavior, attitudes, perceptions, etc. Questions may be "open" (offering a wide scope in answering) or "closed" (requiring constrained answers within a specific formulation, e.g., placement on a scale, selection from a list, yes/no). Fact-finding surveys, such as background questionnaires, may use open questions and qualitative analysis. Survey research relies on closed questions and statistical analysis.

The success and utility of survey research hinges on the validity of the questionnaires and other survey instruments: that they do measure or capture what they intend to, and that what they measure or capture represents the construct under consideration. High-quality survey research makes a substantial planning investment, working carefully on research designs (strategy, constructs and operationalization), validating questionnaires and other survey instruments through pilot studies, and designing the sample. Reliability is increased through the use of sets of questions, which minimize the impact of wording.

The utility of survey research is enhanced by combination with other methods, such as observation and interview, which provide depth and additional perspectives.

Table 15: Survey research tradeoffs

Good for:	Obtaining consistent profiles of the characteristics of a population in terms of the constructs under scrutiny. Allows systematic, generalizable investigation of associations among variables in a social context, when controlled laboratory experiments are not feasible. Can encompass affective, social, and cognitive factors.
Bad for:	Exposing processes. Exposing factors not considered in the instrument design.
Kind of evidence:	Quantitative measures and statistical analysis.
Cost of planning:	High. The success and utility of surveys relies on the design, questionnaire preparation, and pilot testing.
Cost of data collection:	Medium, depending on the extent of the survey.
Cost of analysis:	High, given the potential quantity of data..

Controlled experiments

Experiments are the systematic manipulation of variables under controlled conditions, in order to test hypotheses generated from theories. Hence experimentation is theory-driven, and is characterized by:

- a setting controlled by the researcher
- systematic selection of a representative sample of subjects, and assignment to treatment conditions
- the manipulation of one or more independent variables, in order to observe their effect on the dependent variables

For effective experimentation, the researcher requires control of variables: of the independent variables (those being manipulated by the researcher), and of all intervening variables that might affect the dependent variables (those expected to be affected by the manipulation of the independent variables). The internal validity of an experiment depends on the chain of inference between the hypothesis and the conclusion. An advantage of experimentation is that the high level of control should help reduce threats to validity and hence lend strength to the inferential chain. Another is that it facilitates accumulation of evidence. The external validity reflects how representative the setting and sample are of the target population, and hence the extent to which findings from the experiment can be generalized to other settings, and populations. A disadvantage of experimentation is that the control it exerts reduces the relevance of its findings by stripping away the correspondence between natural events and those in the laboratory. Hence utility of evidence may be limited. Further, there are some factors that cannot be manipulated. Crucial to the utility of experimental findings is the operationalization, the way a construct is 'made usable', in the form of phenomena that can be observed (and measured) in the world.

There are (at least) two classic models of human experimentation which address issues of human variability:

Between-subjects design: Different groups of subjects are assigned to the different treatments. Hence the comparison is between groups or between subjects. The advantage is that subjects come fresh to the treatment; there is no learning or order effect. The disadvantage is the impact of individual differences, which may skew variability in the study.

Within-subjects or repeated measures design: The same subjects are used for all experimental treatments. Hence the comparison is within the same group of subjects. Each subject is measured repeatedly, for each treatment, hence the name "repeated measures". The advantage is that individual differences are equalized across the conditions. The disadvantage is the potential for "order effects" or "learning effects" (variations in performance due to the order in which treatments are experienced)

Table 16: Experiment tradeoffs

Good for:	Control, statistical analysis.
Bad for:	Questions that aren't precise enough yet. Experiments involving human beings are problematic, because people are not fully controllable—it is impossible to eliminate all individual variability. Highly controlled experiments may not have sufficient richness for compelling generalization to real-world settings.
Kind of evidence:	Quantitative data reflecting performance.
Cost of planning:	High.
Cost of data collection:	Can be low, and is related to the number of subjects.
Cost of analysis:	Low to medium, depending on the breadth of the statistical analysis.

Analysis

Like data collection techniques, analysis techniques have purposes which they suit, costs, and conditions for their application. Which analysis is chosen is shaped by the question to be addressed, and by the evidence sought. But it is also constrained by the data collected: how it is selected and how it is recorded. On the one hand, the nature of the data demands or excludes particular analysis. On the other, the nature of the analysis puts minimum requirements on the data.

Two analysis examples are discussed here, to highlight this inter-relation between question, evidence, data, and analysis. They are outlined in order to indicate both how the question may shape the analysis, and how the way the data is selected and recorded constrains the analyses which may be applied.

Protocol (transcript) analysis

Protocol analysis is a general term for the systematic analysis of transcribed speech from empirical studies (e.g., interviews, "think-aloud" monologues, discussions during activities by pairs or groups). Observation, case studies, interviews, open questions on questionnaires—all amass transcripts or written material which must be analyzed. Analysis can be approached in a variety of ways:

- quantitative (based on what can be quantified through counting or measurement),
- qualitative (based on identification of non-numeric patterns and on interpretation of meaning and usage, possibly pre-defined),
- theory-driven (drawing categories from theory, testing hypotheses derived from theory which predicts the outcome)
- data-driven (the data is examined for emergent patterns; such analyses do not —or can not—anticipate outcomes, but rely on finding what can be found in the data that is collected)
- comprehensive (seeking to characterize all of the collected data)
- vectored (having a particular focus, and seeking only specific phenomena within the data)

Hence, one thing that distinguishes approaches is the focus: what is of interest, and at what level of granularity.

The approaches are not mutually exclusive: they may be combined (subjecting one data set to different analyses) or may be used in sequence (with output of one analysis feeding into the next). For example, a data-driven analysis of interview transcripts may identify emergent categories. Those categories may be used as the basis for a coding scheme, and the data may be analyzed afresh by applying that scheme. Alternatively, the data may be divided into sub-sets, with patterns emerging from an inductive analysis of one set tested through their application to another subset or to the whole data set. The findings of an analysis of one data set may be tested by applying that analysis to a different data set, e.g., data collected later or from a different subject sample.

Regardless of approach, the best analyses keep an "audit trail" between the primary data (the actual utterances) and the coded or analyzed forms, so that contexts and sequences can be re-established or re-examined, as needed. It is wise to let the informants "speak for themselves" and hence to maintain the links between excerpts and conclusions.

Because the analysis of qualitative data is a matter of judgment, the researcher must decide how an utterance or action is to be described. A number of techniques are employed to reduce the subjectivity of the process. For example, all of the data can be encoded independently by more than one researcher, resolving discrepancies through discussion and refinement of the coding scheme until an acceptable level of "cross-coder consistency" is achieved. Coding can be done by researchers external to the project, so that they come "fresh" to the analysis scheme. Alternatively, independent coders can "calibrate" to each other through practice and negotiation, and then work on divisions of the data, subject to spot checks.

Particularly in data-driven analyses, it is advisable to review the entire corpus seeking counter-examples, gaps in the patterns, and other evidence that would suggest an alternative interpretation of the data. An important concept in such a "counter-evidence review" is "significant absence": the absence of a pattern or a phenomenon which one might reasonably expect to see.

What follows is an indication of the variety of analysis strategies that might be adopted.

Table 17: Some Analysis Strategies

Trawling:
> The richest possible data is collected (and usually transcribed). Analysis (which can be qualitative or quantitative) is data-driven, seeking emergent patterns or organizing concepts. The aim is usually to determine what's important in some situation—possibly to find out what the important questions are, for subsequent investigation. One initial trawling technique is to "skim the cream": to mark important or compelling passages.

2-pass analysis:
> Requires a reasonably large corpus of data. Data is subdivided, and one subset is analysed in order to identify emergent patterns, from which a formal analysis scheme is derived. The analysis scheme is then applied to the remaining data (and possibly to all of the data as well).

Pre-determined categories:
> Tasks and a data coding scheme are determined based on theory or on previous studies. New transcripts are analysed in accordance with the scheme. This can transform transcript data into a variety of forms, such as quantitative data, process descriptions, instance collections, etc.

Bottom-up analysis:
> Break data into 'units'; then systematically code and collate the lower-level categories. Group progressively into higher-level, more encompassing aggregates.

Top-down analysis:
> Abstract emergent or organizing concepts from the data. Work down, to create outlines of the data, sorting phenomena within the concept divisions.

Analysis need not rely wholly on human interpretation. Once data is in electronic form, it is amenable to *automated analysis*, again in various forms. The simplest form is mechanistic counts, for example of occurrences of words or phrases. But computational linguistics affords a wealth of techniques for characterizing texts. And, again, techniques may be combined. For example, an initial manual coding of features can be augmented by application of automated analysis to the coded data.

These approaches have been described in as "generic" a form as possible, in order to reveal some basic analysis strategies.

Statistical analysis

Statistical techniques, similarly, have purposes, costs, and conditions for their application. For example, statistical tests for association or co-relation are familiar in the context of experimental techniques. *Non-parametric* tests are suitable for experiment designs which test only one independent variable. *Parametric tests* can handle experiment designs which vary more than one independent variable and hence which require more complex statistical treatment. Requirements for statistical significance and power determine minimum numbers of subjects, and different statistical tests have different pre-conditions. For example, parametric tests require interval measurement, normal distribution, and homogeneity of variance. Such conditions have implications for the ability of a given technique to address the complexity of human behavior—the requirements for a particular test may be too constraining to fit the purpose of the research question (for example the assumption of homogeneous variance)—and for the ability of a given technique to be applied within the pragmatic constraints on data collection (for example, limitations on the numbers of available subjects may exclude some tests). Hence, the experiment design shapes the analysis through its focus, the nature of the data constrains which analysis technique may be applied, and the minimum requirements of the statistical technique limit which sorts of situations it may address.

Statistical methods can also be applied to quasi- and non-experimental data, sometimes as a test of association, but more often as a descriptive tool. Again, the question and its evidence requirements shape the analysis desired. Data and technique make demands of each other, the nature of the data constraining which techniques may be applied, and the desired techniques setting requirements for the data to be collected.

Summary

Effective research requires methods which generate data relevant to the research question. Our pragmatic approach hinges on formulating the research question in a way that encompasses not just *what* is asked, but *for what purpose*—and hence establishes what sort of evidence is fit and sufficient to address the question. These point the way to the choice of method. The "what" suggests the sort of data required and hence the sort of method needed, and the "for what purpose" influences the choices about how the method will be applied, in order to maximize utility within the constraints of cost. Study design is a matter of tradeoffs, between richness, resolution and costs; among the costs of different stages of design, implementation, execution and analysis; and among resources (such as numbers of subjects, amount and richness of data, time, and equipment) constrained within a budget.

7

Replicate and Generalize Across Studies

In order to contribute usefully to the discourse, our research findings must be valid, relevant, and important. These qualities are the drivers for the attention to replication and generalization. We need to establish that our findings and conclusions are 'true', that they are neither chance findings nor distortions. One mechanism for doing so is to expose the work to validation—to *replication* or *repetition* and investigation—by others. We need to establish that our questions are significant, and that our findings address those questions usefully. Hence, we hope that the findings *generalize*, that they apply beyond our particular study to reveal some underlying 'truth' applicable to a larger population, set of tasks, or context. We also need to clarify how our findings are bounded—and also what the limits are of the theory that explicates them.

Replication and repetition

Replication and repetition are means for testing validity, in terms of the reliability and robustness of the findings. Replication and repetition are closely related. 'Replication' is the 'verbatim' reproduction of a study by another researcher, that is, using the same protocol under the same conditions. Replication tests how 'reliable' the findings are, that is, how consistent the outcomes of a given study will be given repetition by different researchers, at different times, with a different sample of the same population. Reliability contributes to the strength of evidence. We seek replication in controlled experimentation.

Replication is not necessarily feasible in educational research, which is set in a complex and dynamic social environment that may defy reproduction of conditions. Hence we seek 'repetition' in studies other than experiments, reproduction by another researcher using the same protocol under similar conditions in a similar setting, e.g., moving it from one classroom to a similar classroom with a similar learning context. Repetition tests reliability and also, because of the small differences in context and setting, the 'robustness' of the findings. That is, repetition can show how consistent the outcomes of a given study are across different related tasks, across different environments, across different related contexts. Repetition also exposes study design and conduct to the scrutiny of more minds, and hence puts the inference chain to the test and may help to draw out alternative interpretations.

Generalization and representativeness

Repetition, in offering an indication of consistency of outcomes across slightly different conditions, may help us understand how well findings generalize—or help to indicate a margin of error and to establish what limits might apply to the findings. It is part of the nature of empirical study that we investigate a particular example in the hope that it represents a more general phenomenon, and in the hope that any understanding we derive from the particular may extend to the general phenomenon. In seeking to generalize, we must also seek the boundaries of the generalization and understand that it encompasses some margin of error.

Empirical study is characterized by selection: the selection of subjects, of tasks, of time, of setting, of data collection. Every time a selection is made, its ability to represent what it is selected from (whether population, repertoire of activities, environment, etc.) must be questioned. Representativeness is the key to generalization: if the study is representative, then its outcomes can be generalized to the greater population, to other settings, and so on. The particularity of a research outcome, that is, its lack of representativeness, constrains its utility in the research discourse.

Selection of samples

If we want research to be representative, then we must attend to how we make our selections, to how we 'sample' from the world in order to focus an investigation. We usually use the term 'sample' as shorthand for 'sample of subjects'; it usually refers to a selection of people intended to represent a defined population. But sample may equally refer to a defined population (or set) of artifacts, events, or tasks. For each, the representativeness of the selection must be considered.

The first step in sample selection is the characterization of the population which the sample is to represent—the population to which the results of the research are meant to generalize. The characteristics of the population relevant to the phenomenon of interest must be identified, in order to consider *in what ways* the sample must be representative. (The catch is that this may be difficult to do in advance—the phenomenon might be influenced by population factors you may not consider to be relevant.)

Another step is to decide how large the sample must be. A good 'rule of thumb' is to use the biggest sample one can afford and obtain. This is true of tasks and artifacts, as well as of subjects. Early consideration of the analysis strategy will influence this decision: statistical power depends on sample size, and some statistical treatments require minimum sample sizes. Qualitative analysis is expensive and time-consuming and may suggest depth of treatment rather than breadth in sample selection. Hence the evidence requirement will influence sample size.

There are a number of indicators for a large sample size, such as:
- requirement for a high level of statistical significance (the probability that a result is not due to chance), statistical power (the probability that, if an effect exists, it will be found), or both
- likelihood of high attrition rate
- need to sub-divide groups
- many uncontrolled variables
- likelihood that effect sizes will be small
- population is highly heterogeneous with respect to the variables being studied

In short, a large sample is called for when the likelihood of drawing wrong conclusions from a small sample is high. On the other hand, sample selection can also be matter of diminishing returns. It is worth considering what utility will be gained by, say, doubling a sample. There are times when a small sample will do as well as a large one—it is a matter of fitness for purpose, of the evidence requirements of the research.

Strategies for sample selection may be random or non-random, and may be based on the individual or on groups (that is, sub-groups of the population). In this context, it is important to distinguish between *random* samples (in which all individuals in a population have an equal and independent chance of being selected) and *arbitrary* samples (in which the selection is made on some basis notionally irrelevant to the study; 'any one will do'). The two are not equivalent and have different implications for representativeness. Further, there are *non-random* samples, selected on a basis intended to maximize representativeness of the sample for the purpose of the study. Some general methods are indicated here:

Sampling methods
Simple random sampling:
All individuals in a population have *equal* and *independent* chance of being selected. Entails a measurable degree of uncertainty.

Systematic sampling:
Devise a procedure for selecting every *nth* member of a given list of members of the population.

Stratified sampling:
Assure that subgroups in the population will be represented in proportion to the numbers in the population; select randomly from within the subgroups.

Cluster sampling:
The unit is not an individual but a naturally-occurring group; all members of randomly selected groups are included.

volunteer sampling:
Subjects select themselves. *Note:* Volunteers have been shown to differ from non-volunteers in important ways; therefore, use of volunteers constrains generalization.

Empirical study design, particularly in the context of CS education research, is not a pure exercise, but a pragmatic one, in which factors such as access, cost, and ethics can put sharp constraints on design decisions. The opportunistic nature of much research (e.g., times of transition, briefly available resources) means that practical decisions often intrude. However, opportunism can jeopardize representativeness, and many of the common errors in selecting participants for a study arise from practical compromise, for example:

- selecting people because they're available and appropriate sampling is not convenient;
- selecting participants who are not in an appropriate population;
- using volunteers but failing to ascertain how they may differ from non-volunteers on crucial characteristics or abilities;
- selecting a sample that does not provide for attrition and may be too small by the end of the study.

It is essential, therefore, to consider the limitations attendant on such decisions, and to consider their implications for the value of the evidence gathered. There is no utility in selecting a sample by a method which fails to meet the needs of the research design.

Validity

Validity is the extent to which an account accurately represents the phenomenon to which it refers. More generally, validity is the ability of the research to provide accurate and credible conclusions, building on evidence that is sufficient to warrant the interpretation made. The validity of research is established (or threatened) at many levels, and it affects the value of the results, their representativeness, and the legitimacy of generalization from them.

Table 18: Types of validity (See also: (Campbell & Stanley, 1963))

internal validity:	e.g.	**construct validity:**
addresses how consistently similar results can be obtained for these subjects, for this setting, using these techniques; addresses the quality of inference and conclusions within the study	→	whether the constructs related to a phenomenon are valid, whether the operationalisation (the mapping of the construct onto manifestations in the world) is valid, the comparability of that operationalisation with other studies of the same construct
		validity of measures:
	→	whether measures measure what they claim to, and whether they do so reliably
external validity:		**ecological validity:**
addresses whether the study provides a true reflection of the phenomenon as it occurs the world, hence the generaliseability of the conclusions to other times, settings, and populations	→	whether the setting is representative of settings of the same type and of such settings in the world, and hence whether the findings within one setting are generaliseable to other similar settings
		population validity:
	→	whether the sample is representative of the greater population to which the results are generalized

Bias

Bias threatens the validity of research.

Consider a laboratory experiment that compared two solvents. The setting is controlled: it is a 'fair test', with identical environment, materials, and protocol for the two conditions. Temperature, the nature and amount of material to be dissolved, the application of the solvents – all are identical. And all are arguably representative of natural environmental conditions for the task. One solvent is demonstrably more effective than the other. What conclusions might one draw, how convincing would they be, and how safe would they be to generalize?

But what if the laboratory were also a television studio, and the two solvents were dishwashing detergents. Advertising product comparisons are presented as controlled experiments, but do you consider them to be 'fair tests'? Advertising standards require that the control of variables in product tests be genuine. But they allow the control—the choice of temperature, nature and amount of material to be dissolved, and means of application of the detergents—to be optimized for one of the detergents.

In product comparisons, the comparison is controlled, and the results are reliable, but bias is built into that control. Now, what if the bias were not intentional? Might the appearance of control mask the limitations of the result, and their implications for restrictions on the conclusion?

'Bias' is when things creep in unnoticed to corrupt the evidence. It is the distortion of results due to factors that have not been taken into consideration, e.g.,

- extraneous or latent influences
- unrecognized conflated variables
- selectivity in a sample which renders it unrepresentative

The very act of experimenting introduces the potential for bias. This is referred to as the Heisenberg principle: you can't observe without influencing what you're observing. The fact of observing phenomena changes them.

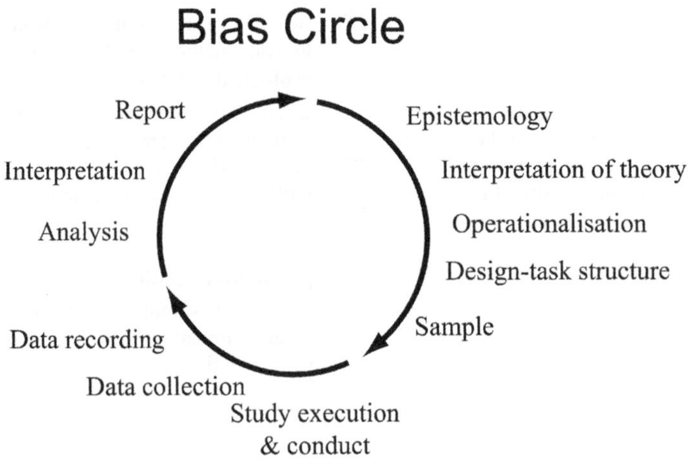

Bias Circle

Figure 6 : Bias Circle. After James Powell (Powell, 1998)

Bias can creep in at any point in research, from the earliest planning through each reasoning and implementation step, through execution, data collection, analysis, and even reporting. In CS education research, fallible, variable humans are both the subjects and the instruments of research, providing multiple opportunities for error and distortion. Rigor demands vigilance against bias, with implications for the design of empirical studies, and for the design and execution of data collection. There are 'dangers around every corner'.

Dangers in operationalisation

A crucial link in the chain of inference is 'operationalisation', linking the concept or construct of interest to an observable indicator—to something that can be investigated empirically. The construct is mapped onto one or more manifestations in the world, things that can be observed, recorded, and ultimately measured in some way. The validity of the study rests on that operationalisation, on that mapping from construct to observable phenomenon to measure. If the reasoning that associates the measure with the construct is faulty, then the data may be irrelevant or misleading.

Operationalisation is important, too, in the accumulation of evidence. Not only is the construct mapped onto some manifestation in the world, but also the mappings applied in different studies must be compared. Is the construct interpreted in the same way? Are the manifestations comparable? Are the measures applied to the manifestations actually measuring the same thing? How well do the measures reflect the manifestation, and how well does the manifestation represent the construct?

The difficulties of finding relevant measures are many:

- Hard to find a measure.
- Hard to be sure it measures what's wanted.
- Hard to be sure it reflects enough of the story.

We've already discussed the difficulty of achieving precision and retaining relevance, characterized as 'sand through the fingers'.

Time and error measures are often used. But what is the meaning of time? Typically, the measure is of performance time, but 'time off task' or 'fiddling time' apparently spent in distraction tasks such as tidying or playing may have a bearing on performance, and they are difficult to measure. What is the meaning of error? Experts tend to make more errors than journeymen (experienced non-experts), but their overall performance is better, because they are better able to recognize and correct their errors, and journeymen expend more time and effort fending off error during initial generation. Time and error are accessible, but they may not be able to account for human perceptions and behavior of interest. For example, motivation can lead people to spend disproportionate time on a task, and yet to perceive it as quick.

Measures are shorthand, a compact expression or reflection of a phenomenon. But they're often *not* the phenomenon—the measure is typically a simplification. Some things are hard to measure, to quantify. For example, making continuous phenomena discrete can distort them. Experimental techniques have us focused on surface features, and the quest for measures can distract us from what is relevant with what is readily measured—sometimes we need other techniques to investigate deeper issues, before we can seek relevant measures.

Dangers in interpretation

The difference between 'data' and 'evidence' is interpretation. Evidence is data, plus the meaning we ascribe to it. Therefore, our reasoning about data is crucial, and it must take into account a variety of dangers in interpretation. There are many dangers, but some are more common than others: selectivity, flaws in reasoning, failing to make alternative accounts, falsely comparing heterogeneous evidence and failing to distinguish "frame of reference".

Selectivity

The danger of selectivity in interpretation is that any description we make of our observations, phenomena, processes, etc., whether to ourselves or in print, is selective. Through the process of research we gather data, but the phenomenon of interest is always more general than the data we choose to collect. Any measure we use to characterize and compare our results is shorthand for some feature of the world.

Flaws in the reasoning chain

The relationships between the phenomena of interest, the research design which aims to capture information about them, the constructs by which we describe them, the ways we operationalise those constructs, and the measures that capture them, are linked by a chain of reasoning, expressed through a chain of argument. Weakness in that chain potentially impairs the relevance and value of the data.

Alternative accounts

Data often admits more than one interpretation, more than one account of its meaning. Alternative accounts should be sought and given due consideration. If possible, alternative accounts should be investigated empirically to establish if they are valid. Alternatives should be ruled out systematically.

Comparing heterogeneous evidence

Some of the most informative studies combine techniques. But one burden of multi-method research is the difficulty of aggregating or comparing heterogeneous evidence. False similarity can lead to false conclusions. Dangers of interpretation are exacerbated when we are reasoning across a number of studies, or across a variety of data.

An exaggerated illustration of the danger is the way evidence was used in a popular TV programme that addressed controversial issues. The program presented heterogeneous evidence in a progression that tended to lead the viewer to draw false conclusions:

Progression	**Fictional illustration**
0. identify a controversial issue	There is concern that fluoride in the water makes us crazy.
1. report the results of a survey (the usual stratified sample format)	Do you think it reasonable to believe that fluoride in the water may have unexpected side-effects?
2. extract the interesting statistic	65% of the population thinks that fluoride may have unexpected side-effects.
3. find a couple of extreme cases and interview them	Jonny and Jimmy think that fluoride made them crazy; let's talk to Johnny and Jimmy and see how strangely they behave
4. add a dose of authority by interviewing scientists	Yes, some studies have been conducted in Europe into the side-effects of fluoride.
5. draw an unsupported causal inference	With such wide-spread concern about fluoride making people crazy...
(6. create panic)	People stop letting their children drink tap water.

Frame of reference differences

Borrowing methods without understanding the disciplinary, methodological, and conceptual framework is dangerous. For example, software engineering is task-oriented, whereas psychology of programming is human-oriented. Both use tasks, and indeed may investigate comparable tasks, but their interpretation may differ because of the disciplinary orientation. Frame of reference differences can distort data collection and lead to specious conclusions.

For example, "A question from [a] language development test instructs the child to choose the 'animal that can fly' from a bird, an elephant, and a dog. The correct answer (obviously) is the bird. Many first grade children, though, chose the elephant along with the bird as a response to the question." (Mehan, 1973)

Any child familiar with the 'Dumbo' film featuring a flying elephant might answer in this way. Test materials do not always have the same meaning for the tester and the subject, i.e., test scoring is interpretive.

There is danger in taking things out of context, and hence losing the original frame of reference. One classic example is 'seven plus or minus two', a limit on working memory established by George Miller in some classic psychology experiments (Miller, 1956). HCI designers have taken the finding up and applied it to interface design, using it as a limit on the number of items in a menu. However, selecting from a menu requires recognition, not recall; the finding is irrelevant to the application.

Danger in naïve appeals to scientific method

One artifact of the dominance of the 'scientific method shorthand'—of the appeal to method without a sufficient perspective on evidence—is a confusion of form with rigor. In fact, naïve approaches to scientific method can produce misleading or false insights. There is no hope of achieving the precision required for controlled experimentation before one understands what the question is, and what evidence is required to address it, and hence what constraints, simplifications, and trade-offs are acceptable for the purpose.

Among the dangers of 'premature experimentation' are:

- ill-formed hypothesis (hence lack of precision, confirmatory bias, danger of uninformative results)
- lack of control (don't know which variables are likely to be important, and hence which to control for)
- uncertain operationalisation (the relationship between the constructs being examined and the particular variables under observation is not established; are the manifestations in the world true reflections of the phenomenon of interest?)
- inadequate measures (the measures are insufficient to capture what they're meant to capture)
- inappropriate expectation (inappropriately seeking proof or conclusive evidence)

The consequences are spurious data, flawed analysis, and false conclusions.

Well-designed experiments are a powerful research tool. 'Scientific method' achieved dominance for good reason. But the 'if I find the right experiment I can do a statistical proof' model of empirical study design is often a case of 'trying to run before one can walk'. In a theory-scarce domain, one needs enough disciplined observation to provide a reasonable basis for identifying important factors and relationships, in order to distil well-founded conjectures (pre-theories?), from which one can generate the sort of precise hypotheses which are worth the cost of experimentation. Premature experimentation can narrow the focus too soon, and miss important phenomena entirely.

The definitive experiment clearly has its place in theory validation. The accumulation and valuation of evidence through a variety of methods is the preparation for theory generation, the obvious prerequisite for theory validation. The definitive experiment is a fine aspiration, but it is perhaps the wrong mechanism for CS education research, when what we need is better questions.

Danger in naïve appeals to metrics and statistics

> The measurement of the 100-yard dash is trivial… Measurement of intellectual artifacts is in its infancy (Curtis, 2000)

Beside the 'scientific method shorthand' walks an uncritical veneration for metrics "Numbers are good. Numbers are objective.", for numerical data, and for statistical analysis. The 'method of science' focuses our attention on questions that can be addressed empirically. The shorthand confuses that with 'what can be measured' or 'what can be addressed experimentally', hence potentially overlooking crucial factors and phenomena. The dangers here are captured in the McNamara Fallacy:

> The first step is to measure whatever can be easily measured. This is OK as far as it goes. The second step is to disregard that which can't be easily measured or to give it an arbitrary quantitative value. This is artificial and misleading. The third step is to presume that what can't be measured easily really isn't important. This is blindness. The fourth step is to say that what can't be easily measured really doesn't exist. This is suicide. (Handy, 1995)

Well-founded, valid metrics are powerful instruments. The key is to find a measure for what is important, rather than to make important what is measurable. How good is the evidence provided by a given metric? From it flows a series of related questions: What does the metric measure? How reliably does it measure it? How does what it measures relate to what we want to know? What *doesn't* it measure that might be important? The power of numbers (or words) in capturing phenomena lies in the validity of the measures (or constructs), in the chain that connects question to operationalisation to data to interpretation. Measures are context-dependent.

All too many studies simply measure the wrong thing. An example comes from software visualization. A researcher had devoted considerable energy to developing a debugging tool, applying a cocktail of metrics in order to select the most complex segments of code. Unfortunately, the work overlooked a pertinent characteristic of programmer behavior: programmers focus their analytic skills, tools, and time on the complex segments during development. Hence, the killer bug is more often in the simple code, the bits that programmers take for granted while they are focusing on the complex code they're worried about.

Researchers often confuse form with rigor not just in their data collection, but also in their analysis—in their appeals to statistics. Statistics 'feel' precise, but that doesn't mean that they are. From statistics, people hope to gain:

- rigor,
- a 'conclusive' demonstration of an effect,
- objectivity.

But, as Huff phrased it: "A difference is a difference only if it makes a difference." (Huff, 1954) Application of statistics without sufficient statistical insight can be meaningless or misleading.

Hence, we offer some cautions against common errors in statistical argument (drawn from Huff):

- Statistics work best in simple cases.
- In a statistical analysis, notions of 'trends' and 'influences' are meaningless if they are not supported by statistically significant results.
- In assessing the strength of statistical evidence, we must consider not just result of the test, but also the levels of significance (the probability that a result is not due to chance) and power (the probability that, if an effect exists, it will be found).
- The failure of data to pass a statistical test doesn't necessarily mean that the effect doesn't exist, only that it wasn't detected in this sample.
- An association between two factors is not proof that one has caused the other. Co-variation often reflects influence from a third factor.
- It is dangerous to infer beyond the data.
- A correlation may be real and based on real cause-and-effect — and still have little utility in addressing the research question.
- "The trend-to-now" may be a fact, but the future trend represents no more than an educated guess.

Tools are as good as the use we make of them. At their best, statistics are an incisive research tool (or collection of tools) that can be used in a variety of ways, e.g.:

- to describe,
- to compare,
- to detect patterns or relationships.

Nevertheless, their status is subject to interpretation: "Statisticians believe that the validity of the statistics can be proven mathematically; whereas mathematicians believe that the validity of statistics can be proven empirically."

Pilot studies

Pilot studies are the first defense against oversight (or stupidity) and the bias it may invite. They help to establish credibility, feasibility, and comprehensibility in advance of the data collection. A good pilot study provides a chance to debug the protocol, to expose frame of reference problems, to test the analysis on genuine data. It can expose design flaws, hidden assumptions, and unexpected problems.

So what makes a good pilot study? It must be a genuine 'dress rehearsal', using the full protocol with subjects representative of target population. Every aspect of the study must be tested out beforehand. The protocol, instruments, and procedures must be tried out, debugged, and tried out again until it is clear that they will work as intended, and that they will generate data which will be pertinent and amenable to analysis. Pilot studies are expensive of time and resources, but the consequences of inadequate testing are likely to be even more expensive. Short-cuts can be catastrophic.

It is crucial that the sample used for the pilot studies be representative of the target population. For example, British academics cannot be taken as representative of European academics (a short-cut that cut one of us short); they may have significantly different interpretations of taken-for-granted terminology. It is also important to pilot the analysis; working back from the analysis can reveal fundamental inadequacies in the study design. The data needs of the statistical tests may expose shortcomings in the data collection. Working back from the analysis may expose gaps in the chain of inference. Better to spot them early than to collect inadequate or irrelevant data.

Accumulation of evidence

Replication is one way of testing the strength of evidence—and potentially of contributing to its strength. Repetition is another, with the additional potential to extend the evidence by accumulating related findings from comparable but differing studies. A condition for replication or repetition is that the study be made accessible, that its definitions, protocols, links to theory, reasoning, and reporting be thorough, accurate, and public. This is also a condition for accumulation of evidence across a number of studies; full access is necessary for the assessment of the relatedness and comparability of constructs and findings.

One of the advantages of standard procedures (of adherence to form) is that it facilitates accumulation of *comparable* data and evidence. Those operating within a given set of standards share epistemology, terminology, conceptual frames, ways of reasoning, ways of reporting, and even assumptions, and this allows them to think about and compare each other's work readily. They get to compare 'apples to apples'.

Hence, one of the burdens of a triangulation approach is to accommodate heterogeneous data, somehow rendering it into comparable forms, or finding means of recognizing regularities. In other words, the challenge is to connect variables among studies so that inferences can be made with increased realism and increased control. For example, field experiments can counter-balance laboratory experiments, if they address the same constructs interpreted in comparable ways—or if the two share enough essential features to be similar.

Any comparison that is made among heterogeneous data must take into account the way that data is colored by how it was collected and interpreted: by the epistemology and disciplinary traditions influencing the casting of the question and the study design, by the assumptions and limitations that attach to the conceptual frames employed in the data's interpretation, by the selections made in the operationalisation and instrumentation, and by the selections and simplifications employed in the description and report. In comparing 'apples with oranges', we need to find a means of reasoning about fruit, while maintaining awareness of the particularities and differences.

There are many issues to consider in making sense across studies:

- *Terminology*: are words used to mean the same things, with the same granularity?
- *Conventions/standards*: what is implied by the conventions and standards observed in the different studies; is some data or reasoning excluded by one and not the other? are there differences in the standards of reporting that may have consequences for the completeness of the accounts? what is considered to be acceptable practice?
- *Assumptions*: are different assumptions implicit in the techniques applied or the theories brought to bear?
- *Conceptual frames/ ways of reasoning*: what assumptions are implicit in the conceptual frames? Are the levels of granularity and abstraction comparable and is the interpretation of concepts or constructs comparable? do differences in reasoning about data lead to differences in legitimacy?
- *Time*: might the effects of time (history, changes over time, fluctuations, patterns or variations in phases) influence the quality of the evidence?

Time is a key issue, often overlooked. Given human memory and psychology, the impact of time—or rather of limitations or considerations associated with time, our perception of it, and the way our perception of time influences our interaction with the world—can be profound in CS education research. For example:

- Initial use does not necessarily generalize to evolved use.
- Single use does not necessarily generalize to repeated use.
- Time is reflected in sequences, processes, antecedents, and context.
- History may have an impact on current phenomena.
- Phenomena may change over time.
- Phenomena (patterns, variations) may occur in phases or have periodic fluctuations.

The focus provided by theory makes it natural to pursue cumulative research. But, in the absence of well-founded theory, attention to the accumulation of evidence contributes to a pragmatic approach to theory-building and theory use. In either case, theory (in the role of the driver, or the goal) provides a focus, making accumulation easier. Accumulation of evidence over a number of studies provides a means of addressing the difficulties of achieving a critical mass of work on a given topic. Attention to accumulation mitigates against isolated and esoteric studies.

The need for honesty

With so many vectors of bias and threats to validity, vigilance is a constant necessity. But so is honesty. The impact of evidence in the discourse depends on people's ability to assess its strength. Good evidence presentation requires clear description of data collection and analysis, an explicit account of the chain of reasoning from study design through data interpretation to conclusions, and an assessment of the reliability and margin of error. Through honest reporting, evidence is exposed to scrutiny, to test and possible falsification.

8

Disclose Research to Encourage Professional Scrutiny and Critique

CS education research is …

As we view—and practice—it, CS education research is not just "scientific method", nor is it confined to the natural world. It borrows from other areas and traditions, in terms of theory, method and approach. We adopt what we term "method of science", a principled and rigorous articulation of observation and explanation.

Which is almost, but not quite, enough. Because science is a discourse, and articulation is chiefly about reading and writing.

Reading is important, because it helps direct research purposefully, providing others' work to build on, indicating which avenues to avoid, showing where contributions are needed. We are fuelled by our scrutiny, critique, and use of others' work. If we don't know what others have done, we stand a good chance of wasting our time by "re-inventing the wheel", unknowingly re-creating work that makes no contribution to knowledge. Whether we use others' work to provide situation and context for our own, or, more closely, as a study to replicate or generalize from, we owe the researchers whose work we use a duty of care, and should practice basic academic skills of reference and report. Naturally, this means giving proper credit. It also entails ensuring that we use others' ideas and work as the originators intended, and not for what we would like them to be or for what we would like them to say.

Writing is important, because otherwise our work is invisible and unscrutinised. We can pose significant questions, choose appropriate methods, operationalise them scrupulously avoiding all possible bias, to uncover evidence which has an impeccable chain of inference. But if we don't then write about it, we might as well not have bothered.

The discourse puts obligations on what we write and how we write it. Research papers are not just telling a story or making a report: they must provide an audit trail of the work and thought that lead to our claims and conclusions. In this way our work can be scrutinized (examined for accuracy) and critiqued (probed for weakness) by our colleagues and peers. If our work is good, then we can expect to be read by others, and perhaps used to situate their own work, or perhaps be replicated by them. We owe them a duty of care to be honest in the framing, situation, conduct, and reporting of our work.

Reading and writing together are about "joining the discourse"

References

Abowd, G. D. (1999). Classroom 2000: An Experiment with the Instrumentation of a Living Educational Environment. *IBM Systems Journal Special Issue on Pervasive Computing, 38*(4), 508-530.

Agre, P. E. (1997). *Computation and Human Experience*. Cambridge: Cambridge University Press.

Anderson, R. J., Anderson, R., VanDeGrift, T., Wolfman, S., & Yashuhara, K. (2003). *Promoting Interaction in Large Classes with Computer -Mediated Feedback*. Paper presented at the Computer Supported Collaborative Learning, Bergen, Norway.

Astrachan, O. (1998). *Concrete teaching hooks and props as instructional technology*. Paper presented at the ITiCSE, Dublin.

Astrachan, O., Wilkes, J., & Smith, R. (1997). *Apprenticeship Learning in CS2*. Paper presented at the 28th SIGCSE Technical Symposium on Computer Science Education, San Jose, CA.

Bannon, L. (2000). *Borrow or steal? Using Multidisciplinary Approaches in Empirical Software Engineering Research*. Paper presented at the International Conference on Software Engineering, Limerick, Ireland.

Ben-Ari, M. (2001). Constructivism in Computer Science Education. *Journal of Computers in Mathematics and Science Teaching, 20*(1), 45-73.

Bloom, B. S. (1956). *Taxonomy of educational objectives; the classification of educational goals* (1st ed.). New York,: David McKay.

Brown, A. L. (1992). Design Experiments: Theoretical and Methodological Challenges in Creating Complex Interventions in Classroom Settings. *The Journal of the Learning Sciences, 2*(2), 141-178.

Brown, A. L., & Palincsar, A. S. (1989). Guided, co-operative learning and individual knowledge acquisition. In L. B. Resnick (Ed.), *Knowing, learning and instruction: Essays in honor of Robert Glaser*. Hillsdale NJ: Lawrence Erlbaum Associates.

Bruner, J. (1960). *The Process of Education*. Cambridge MS: Harvard University Press.

Campbell, D. T., & Stanley, J. C. (1963). Experimental and quasi-experimental designs for research. In N. L. Gage (Ed.), *Handbook of Research on Teaching* (pp. 5-6). Chicago: Rand McNallly.

Csikszentmihalyi, M. (1991). *Flow - the Psychology of Happiness*: Rider.

Curtis, B. (2000, 5 June). *Borrow or steal? Using Multidisciplinary Approaches in Empirical Software Engineering Research (Keynote talk)*. Paper presented at the International Conference on Software Engineering, Limerick, Ireland.

Dourish, P. (2001). *Where the Action Is: The Foundations of Embodied Interaction*. Cambridge MA: MIT Press.

Edwards, P. N. (1996). *The Closed World: Computers and the Politics of Discourse in Cold War America*. Cambridge MA: MIT Press.

Eisenstadt, M. (1993). *Tales of Debugging from the Front Lines*. Paper presented at the Empirical Studies of Programmers.

Engestrom, Y., Miettienen, R., & Punamaki, R.-L. (Eds.). (1999). *Perspectives on Activity Theory*. Cambridge, UK: Cambridge University Press.

Entwistle, N. J., & Tait, H. (1995). *The Revised Approaches to Studying Inventory*. Edinburgh: Centre for Research on Learning and Instruction, University of Edinburgh.

Feynman, R. P. (2001). *The Pleasure of Finding Things Out*. London: Penguin.

Fincher, S. (1999). *What are we doing when we teach programming?* Paper presented at the Frontiers in Education, San Juan, Puerto Rico.

Fincher, S., Petre, M., & Clark, M. (Eds.). (2001). *Computer Science Project Work: Principles and Pragmatics*. London: Springer-Verlag.

Foxley, E. (2003). *Ceilidh*, from http://www.cs.nott.ac.uk/~ceilidh/

Galison, P. L. (1997). *Image and logic : a material culture of microphysics*. Chicago: University of Chicago Press.

Gilmore, D. J., & Green, T. R. G. (1984). Comprehension and the recall of miniature programs. *International Journal of Man-Machine Studies, 21*, 31-48.

Glaser, B. G., & Strauss, A. L. (1967). *The Discovery of Grounded Theory: Strategies for Qualitative Research*. Chicago.

Handy, C. (1995). *The Age of Paradox*. Cambridge, MA: Harvard Business School Press.

Hause, M. L., Almstrum, V. L., Last, M. Z., & Woodroffe, M. R. (2001). *Interaction Factors in Software Development Performace In Distributed Student Groups In Computer Sceince*. Paper presented at the 6th Conference on Innovation and Technology in Computer Science Education, Canterbury, England.

Hempel, C. G., & Oppenheim, P. (1965). Studies in the Logic of Explanation. *Philosophy of Science, 15*, 135-175.

Holmstrom, J. E. (1947). *Records and Research in Engineering and Industrial Science: A guide to the sources, processing and storekeeping of technical knowledge with a chapter on translating* (Second ed.). London: Chapman and Hall Ltd.

Huff, D. (1954). *How to Lie with Statistics*. London: Penguin Books.

Isaac, S., & Michael, W. B. (1989). *Handbook in Research and Evaluation for Education and the Behavioural Sciences*. San Diego, CA: EdiTS Publishers.

Jenner, E. (1798). *An inquiry into the causes and effects of the Variolae Vaccinae, a disease discovered in some of the western counties of England, particularly Gloucestershire, and known by the name of the cow-pox*.

Kember, D. (1995). *Open Learnig Courses for Adults: A Model of Student Progress*. Eaglewood Cliffs, NJ: Educational Technology Publications.

Kolb, D. A. (1984). *Experiential learning: experience as the source of learning and development*. Englewood Cliffs, NJ: Prentice-Hall.

Kolikant, Y. B.-D. (in press). Learning Concurrency as an Entry Point to the Community of CS Practitioners. *Journal of Computers in Mathematics and Science Teaching*.

Kraemer, K. L. (1991). Introduction. In K. L. Kraemer (Ed.), *The Information Systems Research Challenge: Survey Research Methods* (Vol. 3, pp. xiii-xvii): Harvard Business School.

Kuhn, T. S. (1970). *The Structure of Scientific Revolutions*. Chicago: University of Chicago Press.

Lancaster, T., & Culwin, F. (2004). A Comparison of Source Code Plagiarism Detection Engines. *Computer Science Education, 14*(2).

Lave, J., & Wenger, E. (1991). *Situated learning : legitimate peripheral participation*. Cambridge England ; New York: Cambridge University Press.

Linn, M. C., & Clancy, M. J. (1992). The Case for Case Studies of Programming Plans. *Communications of the ACM, 36*(3), 121-132.

Lister, R., & Leaney, J. (2003). *Introductory Programming, criterion-referencing, and Bloom*. Paper presented at the 34th SIGCSE Technical Symposium on Computer Science Education, Reno, NV, USA.

Mantei, M. M. (1989). A discussion of "Small group research in information systems: theory and method". In *The Information Systems Research Challenge: Experiemental Research Methods* (Vol. 2, pp. 89-94): Harvard Business School.

Mason, R. O. (1989). MIS experiments: a pragmatic perspective. In I. Benbasat (Ed.), *The Information Systems Research Challenge: Experimental Research Methods, Vol. 2* (pp. 3-20): Harvard Business School.

Masterman, M. (1970). The Nature of Paradigm. In I. Lakatos & A. Musgrave (Eds.), *Criticism and the Growth of Knowledge*. Cambridge: Cambridge University Press.

McCracken, M., Almstrum, V., Guzdial, M., Hagan, D., Kolikant, Y. B.-D., Laxer, C., et al. (2001). A multi-national, multi-institutional study of assessment of programming skills of first year CS students. *SIGCSE Bulletin, 33*(4), 125-180.

Mehan, H. (1973). Assessing Children's Language Using Abilities. In J. M. Armer & A. D. Grimshaw (Eds.), *Methodological isses in compartative sociological research*. New York, USA: John Wiley and Sons.

Miller, G. A. (1956). The Magical Number Seven, Plus or Minus Two: Some Limits on Our Capacity for Processing Information. *The Psychological Review, 63*, 81-97.

Mislevy, R. J. (2001). *Basic concepts of evidentiary reasoning*, from http://www.education.umd.edu/EDMS/EDMS738_

Myers, B. A. (2001). Using Hand-held Devices and PCs Together. *Communications of the ACM, 44*(11), 34-41.

Myers, I. B. (1998). *MBTI manual : a guide to the development and use of the Myers-Briggs Type indicator* (3rd ed.). Palo Alto, Calif.: Consulting Psychologists Press.

Myers, I. B. (2000). *Introduction to Type: A Description of the Theory and Application of the Myers-Briggs Type Indicator*. Oxford: Oxford Psychologists Press.

Nardi, B. (Ed.). (1996). *Context and Consciousness: Activity Theory and Human-Computer Interaction*. Cambridge, MA: MIT Press.

OECD. (1995). *Background Paper to the OECD Workshop: Sustainable Consumption and Production: Clarifying the Concepts*, from http://www.sustainableliving.org/appen-e.htm

Palincsar, A. S., & Brown, A. L. (1984). Reciprocal teaching of comprehension -fostering and monitoring activities. *Cognition and Instruction, 1*(2), 117-175.

Papert, S. (2003). *Seymour Papert*, from http://www.papert.org/

Perry, W. G. (1981). Cognitive and Ethical Growth: The Making of Meaning. In A. W. Chickering (Ed.), *The Modern American College* (pp. 76-116). San Francisco: Jossey-Bass.

Perry, W. G., & Harvard University. Bureau of Study Counsel. (1970). *Forms of intellectual and ethical development in the college years; a scheme*. New York,: Holt Rinehart and Winston.

Petre, M., Price, B. A., & Carswell, L. (1996, April). *Moving programming teaching onto the Internet: experiences and observations*. Paper presented at the 8th Workshop of the Psychology of Programming Interest Group, Ghent.

Popper, K. (1959). *The Logic of Scientific Discovery*. New York: Basic Books.

Powell, J. (1998). In M. Petre (Ed.).

Research, C. f. A. E. a. S. E. (1995). *Phenomenographic Research: An Annotated Bibliography* (Occasional Paper 95.2). Brisbane, Australia: QUT Publications and Printing.

Schkade, D. A. (1989). Prospects for Experiments Focusing on Individuals in IS Research. In I. Benbasat (Ed.), *The Information Systems Research Challenge: Experimental Research Methods* (Vol. 2, pp. 49-52). Boston, MS: Harvard Business School.

Seger, C. A. (1994). Implicit Learning. *Psychological Bulletin, 115*, 163-196.

Shavelson, R. J., & Towne, L. (Eds.). (2002). *Scientific Research in Education*. Washington DC: National Academy Press.

Sheard, J., Dick, M., Markham, S., Macdonald, I., & Walsh, M. (2002). *Cheating and plagiarism: perceptions and practices of first year IT students*. Paper presented at the Proceedings of the 7th annual conference on Innovation and technology in computer science education, Aarhus, Denmark.

Skinner, B. F. (1938). *The behavior of organisms; an experimental analysis*. New York,: Appleton-Century-Crofts.

Skinner, B. F. (1968). *The technology of teaching*. Englewood Cliffs, N. J.: Prentice-Hall.

Sphorer, J. C., Soloway, E., & Pope, E. (1985). A goal/plan analysis of buggy Pascal programs. *Human-Computer Interaction, 1*, 163-207.

Stokes, D. E. (1997). *Pasteur's Quadrant: Basic Science and Technological Innovation*: The Brookings Institution.

Vygotsky, L. S. (1962). *Thought and Language*. Cambridge MS: MIT Press.

Wenger, E. (1998). *Communities of practice : learning, meaning, and identity*. Cambridge: Cambridge University Press.

Williams, L., & Kessler, R. (2001). Experimenting with Industry's "Pair-Programming" Model in the Computer Science Classroom. *Computer Science Education, 11*(1).

Notes

[1] We do not at all believe that there is lesser value in "practice papers" than in "perspective papers" or "CS education research" papers. But they serve a different

community for a different purpose. The focus of practitioners—educational professionals—is their students, in their classroom. This is almost inevitably located within an institution and an institutional context.

The focus of the researcher is in ideas, experimenters, in theories and their exponents. The community that is interested in these things is almost inevitably distributed—certainly geographically and possibly temporally, as we draw on earlier ideas. Practice papers are an important way to extend exchange within the community of CS education professionals; meetings and conference that support them are a mechanism to expand the context of practice from a single classroom, and hence expand the community of practice from a single institution (*See*: (Wenger, 1998)).

[2] Richard Feynman uses a similar term, "cargo-cult" science in analogy to the behavior of certain remote peoples, who built runways in order to tempt airplanes to land.

[3] Richard Feynman describes an iconic example of the control of variables "… there have been many experiments running rats through all kinds of mazes, and so on— with little clear result. But in 1937 a man named Young did a very interesting one. He had a long corridor with doors all along one side where the rats came in, and doors along the other side where the food was. He wanted to see if be could train the rats to go in at the third door down from where he started them off. No. the rats went immediately to the door where the food had been the time before.

The question was, how did the rats know, because the corridor was so beautifully built and so uniform, that this was the same door as before? Obviously there was something about the door that was different from the other doors. So he painted the doors very carefully, arranging the textures on the faces of the doors exactly the same. Still the rats could tell. Then he thought maybe the rats were smelling the food, so he used chemicals to change the smell after each run. Still the rats could tell. Then he realized the rats might be able to tell by seeing the lights and the arrangement in the laboratory like any commonsense person. So he covered the corridor, and still the rats could tell.

He finally found that they could tell by the way the floor sounded when they ran over it. And he could only fix that by putting his corridor in sand. So he covered one after another of all possible clues and finally was able to fool the rats so that they had to learn to go in the third door. If he relaxed any of his conditions, the rats could tell … I looked into the subsequent history of this research. The next experiment, and the one after that, never referred to Mr Young. They never used any of his criteria of putting the corridor on sand, or being very careful. They just went right on running rats in the same old way, and paid no attention to the great discoveries of Mr Young, and his papers are not referred to, because he didn't discover anything about the rats. In fact, he discovered *all* the things you have to do to discover something a about rats. But not paying attention to experiments like that is a characteristic of Cargo Cult Science" (Feynman, 2001)

[4] There are other well-known and well-explored factors here, of course. Age, culture and preparedness will affect performance. Perhaps more interesting is the idea that the value that we put upon these indicators is extrinsic to what they measure. Because our society values "high IQ" then performance on this scale is more valued than being on the end of other scales of empirical law: being tall, perhaps, or being able to store 12 things in short term memory

[5] It is unusual for any academic discipline to have this strand of tool-building, of the creation of artifacts which is interwoven with the research agenda. However, CS research, too, also has a strong tradition of instrument-building, encompassing simulations, visualizations, algorithm animation, as well as the construction of whole environments. In a CS education research context these are designed to have an effect on the teaching and learning of computing concepts. And, in that context, they should be grounded in theory and refined through empirical study. However it might be that they are a unique contribution from CS education to other disciplinary areas, trade goods of value.

[6] Of course, subsequently, with historical perspective, the value of the trade to both parties can be seen to be different again. We would now feel that the Lenape made a very bad trading deal.

[7] This description taken from: William J. Rapaport *William Perry's Scheme of Intellectual and Ethical Development*,
`http://www.cs.buffalo.edu/~rapaport/perry.positions.html`

Part Two: perspectives and approaches

Introduction

Part One was intended to provide an overview of the field and endeavor of CS education research. Part Two is intended to 'zoom in' on the field, through more detailed exploration of representative topics and literatures in specific areas of CS education research. Part Two (like Part One) is indicative, rather than comprehensive. Each chapter in Part Two is an overview of a particular topic in CS education research written by an expert.

Each chapter is designed to provide, in its own way, a practicing researcher's perspective on what is important within the topic. Each gives entry points into a more focused literature, and exemplars from that body of research. Part Two models discourse, describing work exposed to *professional scrutiny and critique*: here we briefly situate their accounts into the broader context of the six principles explored in Part One.

Link research to relevant theory ...

Firstly, we can see different relationships to theory bases and to reference disciplines from the different contributors. For example, Clear consciously exposes a theoretical approach to CS education research to make it available to others. The chapter is distinctive in rooting the discussion explicitly in an articulated epistemological context, focused through a conceptual lens. The method of presentation demonstrates how critical enquiry research is an "approach to investigation" which constrains questions and methods through a special *chain of reasoning*.

The Rist chapter views acquisition of programming skill through the lens of a single theory, namely schema theory. In contrast, Clancy's chapter is set against a theory base, showing how different selections of theory have been applied to the topic by different researchers. He demonstrates how theory is articulated within the topic.

In his chapter, McCracken takes a single subject—design—and demonstrates how the topic has two camps, which draw on two reference disciplines: one camp inspired by cognitivist constructions of the activity (psychology), and the other camp viewing design as a social activity (sociology). These two traditions bring different research framing and articulation to investigation in this area

Use appropriate methods ...

In contrast, Stasko and Hundhausen use methods as the organising principle for their chapter. They expose the quantity and variety of methods applied in the topic, contrasting the evidence each may produce, and ultimately espousing a multi-method approach. The chapter by Guzdial portrays a body of research dominated by an 'engineering approach' to CS education: conjectures about the nature of student difficulties are addressed by building systems to overcome them.

Accumulation of evidence (to replicate and generalize across studies) ...

Guzdial makes a beautiful exposure of how evidence is accumulated, how it is built up in an area over time, how some research extends other work. He displays graphically how a field builds through accumulation of evidence, identifying 'families' of environments and showing development of ideas both within and across families. Clancy's chapter, too, documents the accumulation of evidence, but the use of accumulation within the field is different, because designs are not directly absorbed and built on, but he shows how the ideas propagate.

The six chapters are different in character, just as the research they are summarising is different in character. Juxtaposed and taken together, they indicate just how varied the perspectives and research approaches can be. Through the overlaps in subject matter among the chapters we hope that the contrast of perspectives will provide insight. We leave it as an exercise for the reader to identify the significant questions that each motivates the areas.

1

Misconceptions and Attitudes that Interfere with Learning to Program

Michael Clancy

Introduction

An instructor of an introductory programming course, on grading exams, notices that a significant number of students are coming up with the same wrong answers; a student arrives at the instructor's office with program bugs relating to language constructs covered a month ago and exercised ever since; a teaching assistant says, "I tell them and tell them, but they still don't get it right!"

The instructor or t.a. might wonder whether the students aren't reading the textbook or not paying attention in lecture. More likely, however, the students *are* trying to learn. They're trying to build understanding of a concept, however, not just from instruction but from experimentation, analogies to other concepts, intuition, and other knowledge. Often inaccurate understanding, represented by systematic *misconceptions[1]*, is the result. Symptoms of misconceptions range from errors involving the details of a given procedure to complete rejection of practices that students are intended to adopt.

Identifying misconceptions and their causes, and devising ways to address them, constitute a significant area of science education research, and there is a considerable literature on how misconceptions and inappropriate attitudes complicate learning. (See National Research Council (1997) for some examples.)

We focus here on research involving misconceptions relating to computer programming.

In this chapter, we start with some background on how misconceptions form. We then survey numerous examples of research revealing programming-related misconceptions caused by inappropriate transfer and confusion about computational models. In most cases, the correspondence between student errors and the underlying misconception is clear, but we note several instances where the connection was not so obvious. We next describe related work involving inappropriate attitudes toward programming. We discuss some way for dealing with misconceptions in programming courses. We conclude by suggesting some directions for research in these areas.

Background

Early misconception-related work included efforts by Brown, Burton, and VanLehn to construct a *diagnostic model* of a student learning subtraction (Brown & Burton, 1978). The model represented subtraction operations as procedures, and proposed that slight perturbations in the procedures were the cause of subtraction "bugs". The model provided a mechanism for explaining *why* a student makes a mistake instead of simply identifying the mistake.

This led to a theory of bug *generation*, which Brown and Van Lehn (1980; VanLehn, 1990) termed *repair theory*. The theory had several components: a formal representation of the correct procedures; a set of "deletion principles" that model miscomprehension of examples and forgetting, and produce a set of *incomplete* procedures; a monitor that detects an *impasse* when an incomplete procedure is applied; a set of *repair heuristics* that propose repairs to handle the impasse; and a set of *critics* to filter out some repairs. The theory drew criticism (Laurillard, 1990; Hennessy, 1990) for its failure to account for a variety of empirical phenomena: bug instability and fragile knowledge (Perkins & Martin, 1986; Hook, Taylor, & Du Boulay, 1990); the propensity of students to hold simultaneous inconsistent models; and real-world conceptualization and semantic knowledge.

Taylor (1990) took a somewhat different view based on *abductive reasoning* as an important learning strategy. An abductive inference finds a hypothesis that explains a situation, determines that no other hypothesis explains it better, and adopt the hypothesis. Explanations generated in this way sometimes keep the learners from noticing their confusion and, when generalized, cause even more confusion. A complication is that beginners are unlikely to test their hypotheses. However, because beginners have no specific domain knowledge, they *must* use abduction or they get nowhere.

Taylor analyzed protocols of novice PROLOG programmers according to how the subjects moved from natural discourse (used between humans) and formal discourse (between humans and computers). Her subjects did not appreciate the differences between the two; she proposed that the resulting inconsistency produced "superbugs" described by Pea (1986). She found that students introduced to PROLOG via declarative programming were more likely to use natural language as a reference point for understanding programming, while students introduced to

PROLOG's procedural semantics were more likely to interpret its execution strategies as more intelligent than they really are.

Eylon and Linn (1988), in a survey of a variety of perspectives on science learning, noted more general features of misconceptions:

> [S]tudents begin their study of science with strongly held conceptions about some phenomena, conflicting ideas about similar phenomena, and little knowledge of other phenomena. ... Studies often report many missing or incorrect links between concepts. ... [S]tudents can hold inconsistent or even contradictory ideas by keeping their knowledge isolated.

Today, *constructivist* research focuses on the links among concepts and how they are built and reorganized to produce learning. Constructivism, according to Smith, diSessa, & Rochelle (1993), is "the view that all learning involves the interpretation of phenomena, situations, and events, including classroom instruction, through the perspective of the learner's existing knowledge". To constructivists, misconceptions naturally occur as part of the transfer and linking process. (National Research Council (2000) provides more details.)

The methodology of exploring misconceptions typically proceeds in two phases: detecting and identifying them, then helping students to revise their understanding. The detection effort sometimes involves analysis of student-submitted code or paper-and-pencil worksheets to guess at what problems students might have. However, misconceptions are built upon the experience of *individual* students; thus the more prevalent approach to data gathering involves interviews of students writing, reading, or debugging code while "thinking out loud".

Over- or undergeneralizing as a cause of misconceptions

We now move to some examples of programming misconceptions that arise from inappropriate transfer—overgeneralizing or undergeneralizing knowledge from other areas.

English
Linguistic transfer is one source of confusion: terms don't always mean the same in English as they do in programming. Working with college students learning Pascal, Soloway and his colleagues noticed (Spohrer & Soloway, 1986; Bonar & Soloway, 1989) mismatches between English and programming vocabulary:
- The word "while" in English implies a continuously active test, as in "*while* the highway stays along the coast, keep following it north." In Pascal, the `while` loop makes one test per iteration. A student who identifies the Pascal `while` with the real-life "while" may expect a loop to terminate *at the moment* the test expression is satisfied.
- The word "then" in English suggests temporal sequence. Applying this meaning to Pascal, one of Soloway's subjects invented a `repeat ... until ... then ...` construct that used the keyword "then" to specify statements executed *after* the loop finished executing.

Pea (1986) observed similar confusion with the word "if". Novice BASIC programmers predicted that an `if` statement was "waiting for" its test to be satisfied; once that happened—again, checked with a continuously active test—the `then` part would be executed.

Scherz, Goldberg, & Fund (1990) report interference of English with coding of predicates in PROLOG, citing the example of students entering a fact such as

```
father (abraham, isaac).
```

into their program to denote that Abraham is the father of Isaac. They will then ask a question

```
?- father (abraham).
```

and expect it to succeed, even though the 2-argument predicate relating Abraham and Isaac is totally unrelated to the 1-argument query. Scherz tested whether the same error occurred with facts that have no natural language connotations, *e.g.*

```
a (b,c).
```

They found that students did not expect a positive answer to the query

```
?- a (b).
```

justifying their suspicion that only when PROLOG clauses look like English sentences do students expect the computer to understand and to reason.

At Berkeley, we have recently noticed similar behavior among students in our introductory Scheme programming course. The dialect of Scheme we use includes "word" and "sentence" data types (following Harvey & Wright (2000)); an empty word is denoted by `""` and an empty sentence by `( )`. Since there are no empty words or sentences in English, students regard them as meaningless. This perception leads students to view the sentence (`mike clancy`) and (`mike "" clancy ""`) as identical.

Mathematical notation

Another common source of overgeneralization is algebraic notation. Several researchers (e.g. Bayman & Mayer (1983); Putnam, Sleeman, Baxter, & Kuspa (1989); Haberman & Kolikant, (2001)) have observed the confusion presented to novices by the assignment statement a = a + 1 in Fortran or a C-based language. Davis, Linn, Mann, & Clancy (1993), investigating misconceptions surrounding the use of parentheses and quotes in Common Lisp, reported that some students thought that the quote mark specifying a literal list could "distribute" onto all the elements of the list.

Previous programming experience

Overgeneralization may result from earlier exposure to different programming languages. Scherz *et al.* (1990) in their work with PROLOG students report interference of previous programming experience with the use of variables, predicates, and arithmetic functions, which work very differently in PROLOG than in more conventional languages. Our Berkeley students' background in Scheme, in which lists are typically compared element by element using the `equal?` predicate, seems to lead them down the garden path of using ==, a very different comparison operator, to compare strings in a C-based language. (Our students are even less ready for C's string comparison predicate, `strcmp`, which returns true when the strings *do not* match.)

Overgeneralizing from examples

There are numerous reports of students constructing incorrect rules by overgeneralizing from examples they've seen. Interviews conducted by Davis *et al.* (1993) of students in an introductory LISP course suggested several such rules:
- all function arguments must be enclosed in parentheses;
- all atomic arguments must be enclosed in parentheses;
- all numeric arguments must be enclosed in parentheses.

In retrospect, it was easy to find examples from lecture or the textbook from which these rules could have been inferred.

Holland, Griffiths, & Woodman (1997) and Fleury (2000) found similar instances of apparent overgeneralization of examples (or lack thereof) in an object-oriented context. Holland observed that examples of classes with only a single instance variable, or in which all instance variables are the same type, might lead students to confuse an object with one of its instance variables or with its class. Fleury's list of overgeneralized rules included the following, for which students had probably not seen any counterexamples:
- two classes can't have matching method names;
- arguments to methods must be numbers;
- the dot operator can only be applied to methods, not instance or class variables.

Modification of correct rules

Confused students may overextend or overrestrict correct rules. In an earlier study (Fleury, 1991), Fleury found that students in an introductory Pascal programming class constructed rules for parameter usage that were slight variations of simplified rules explained in class:

correct rule	*incorrect student rule*
When a procedure references a name that is not declared either among its local variables nor among its parameters, the procedure next searches for a declaration of the name at the next level outside the procedure (static scope).	When a procedure references a name that is not declared either among its local variables nor among its parameters, the procedure next searches for a declaration of the name in the procedure that called it (dynamic scope).
When you want a procedure to change the value of a variable, use a `var` parameter.	When the value of a global variable is changed in a procedure, the new value will not be available to the main program because no `var` parameter passed the value out of the procedure.

Analogy

Finally, there is misapplication of analogy. Du Boulay (1989) cites an example: "A variable is like a box." A box can hold more than one thing; some students think that a variable can as well. Pea (1986) observes that novices view programming—in any language—as analogous to conversing with a human; as a result, they attribute goal-directedness or foresightedness to the program, or assume that the computer will do what they mean (rather than what they tell it to do).

Halasz & Moran (1982), in a paper titled "Analogy Considered Harmful", predict that analogical models of computing are doomed to failure:

An analogical model is, by definition, a partial mapping to the computer system it is supposed to explain. No simple analogical model is sufficient to completely explain the operation of a computer system. Computer systems are too different from familiar, everyday non-computational systems. This forces the use of baroque or multiple analogical models. Furthermore, analogical models explain too much; there are always aspects of the analogical models that are irrelevant to the system being modelled. To make effective use of analogical models, the new user is faced with the confusing task of sorting out the relevant inferences from among the many possible irrelevant or incorrect inferences suggested by the analogy.

A confused computational model as a cause of misconceptions

A high-level language provides procedure and data abstractions that make it a better problem solving tool, but which hide features of the underlying computer from the user. These abstractions, especially if they have no real-life counterparts, can prove quite mysterious to the novice.

Input
For example, input statements are a problem area noted by several researchers (D. Sleeman, Putnam, Baxter, & Kuspa, 1988; Putnam et al., 1989; Bayman & Mayer, 1983; Haberman & Kolikant, 2001). Students wonder where the input data comes from, how it is stored, and how it is made available to the program. Haberman and Kolikant (2001) hypothesized two incorrect models for input processing: a history-driven mechanism that, for a program with several input statements referring to the same variable, preserved all the values ever associated with the variable; and a priority-driven mechanism that activates statements not in sequence but in importance (e.g. an input statement has a higher priority than an assignment statement).

Constructors/destructors
Memory allocation in C++ or Java occurs invisibly. In a junior-level algorithms course where students were competing for fastest-running programs, we noticed that students using C++ seemed to be at a disadvantage compared to their C-using classmates, mainly because they didn't understand C++'s memory management model. Fleury (2000) found a similar misconception in students in an introductory Java course. These students believed that calling a method that sets the values of an object's instance variables could replace a constructor call, ignoring the latter's responsibility for allocating storage for the object.

Recursion
Recursion is another a significant source of misconceptions. Anderson, Pirolli, & Farrell (1988) explain that recursive mental procedures are difficult, perhaps impossible, for humans to execute, and unfamiliar as well. Thus for most people, recursive programming is their first experience with specifying a recursive procedure.

In early work, Kahney (1983) asked 30 college students in a cognitive psychology summer school program to predict the behavior of recursive SOLO program (SOLO is a Logo-like data manipulation language). He noticed that roughly half his subjects had a "looping" model of recursion. That is, they viewed a recursive procedure as a single object rather than a series of new instantiations. Consistent results were found in other subject groups and languages, *e.g.* Logo (Kurland & Pea, 1989) and LISP (Pirolli & Anderson, 1985).

Dicheva and Close (1996) asked children aged 10 to 14 learning Logo to predict the behavior of procedures with an embedded recursive call. They inferred a more detailed categorization of loop models involving the "head block" of statements preceding the recursive call and the "tail block" of the statements following; an example is "the head block is executed as a loop and the tail block isn't executed at all". These models were sub-categorized according to accompanying data flow models. Dicheva and Close also hypothesized a number of misconceptions that would lead to these models. For example, Logo's return-from-procedure command is called stop; an obvious inference is that it stops all the recursive calls of the procedure. Other misconceptions involved data flow, such as beliefs that variables with the same name refer to the same information, or that Logo remembers the value of arguments in the first recursive call for use in subsequent recursive calls.

Execution of PROLOG programs

PROLOG has been taught in introductory programming courses, mostly in Europe, since the early 1980's. However, experience has shown that many students have considerable difficulty using the language. There are a variety of reasons for these problems—we mentioned some of them above—but one is certainly the relatively complex execution model. PROLOG is a *declarative* language of facts and queries, intended to allow the programmer to verify a logical relation without worrying about how its truth was established. The operations of *unification* (establishing consistency of a query, possibly containing variables, with facts in the data base) and *depth-first search* (for instantiations of the variables that satisfy the query) are provided as primitives in PROLOG. These are both powerful and difficult to understand, even for experienced programmers:

> PROLOG's behavioural component is complex, and, because its syntax is noncommittal, learners are tempted to hallucinate onto it whatever they think appropriate, rather than referring to an interpretation based upon underlying domain knowledge. (Taylor, 1990)

> [S]tudents [seem to] view the PROLOG interpreter as a system that desperately searches for a successful unification as it searches for a positive answer to a query. (Van Someren, 1990a)

It is not surprising, then, that students learning PROLOG develop a host of misconceptions. Van Someren (1990b) constructed a variety of "malrules" from correct rules for PROLOG behavior, and found that they explained a significant number of typical student errors. Fung, Brayshaw, Du Boulay, & Elsom-Cook (1990) identified and categorized a large number of student errors and misconceptions.

Masked Misconceptions

Programming knowledge is not just a collection of individual elements, but rather a *system* with complex substructure. Thus the connection between student errors and an embedded misconception that produces them may not be clear. For example, Segal, Ahmad, & Rogers (1992) describe a longitudinal study over the course of two years investigating semicolon errors made by undergraduates learning ALGOL 68. Weekly worksheets were collected from the 100 students in the class, as well as copies of all programs they submitted to the compiler. The data made clear that semicolon errors weren't occurring at random; a large percentage involved omitting semicolons after blocks (one or more statements enclosed in do ... od, the ALGOL 68 equivalent of begin ... end). Students had been told in lecture that

"Given A ; B where A and B are processes, ';' means 'wait until process A has finished and then start process B."

and

"A *process* is anything that processes data."

The culprit, it appeared, was students' misunderstanding of the word "process". They were happy to accept simple statements as processes. Complex structures such as procedure declarations or loops didn't qualify, at least immediately. The moral: A learner's developing understanding of the terms used to define a syntax rule may play a crucial role in understanding that rule.

An example from personal experience involved student errors with parameters of Pascal procedures. In our introductory course in the late 1970's, we started with a week or two of Karel the Robot (Pattis, 1981), an environment programmed in a language like Pascal except for the absence of variables. We moved on to Pascal, covering procedures and functions relatively early (prior to arrays). Each term, we noticed numerous student errors with parameter passing that we were uncertain about how to address. One year, an instructor got carried away and spent half the ten-week term on Karel. His students produced structured Karel programs of several hundred lines and seemed to understand control flow and procedural decomposition well. When they moved to Pascal, however, they were stumped by the idea of a variable. We had taken understanding of variables for granted; clearly this was inadequate, and it immediately became obvious that if students didn't understand how variables worked, they would be equally confused about parameter passing.

Similarly, confusion about recursion may arise from other areas (George, 2000), for instance, misunderstanding of variable updating and computer memory storage, or difficulty with evaluating conditional statements.

Eisenberg, Resnick, & Turbak (1987) report misconceptions about procedures in sixteen MIT students learning to program in Scheme. Scheme procedures are *first-class*; that is, they can not only be called, but can be passed as arguments to and returned as values from other procedures, as well as stored in data structures. For many students, a serious stumbling block to making use of these features is their more simplistic view of procedures as "active manipulators of passive data" and as "incomplete entities that needed 'additional parts' before they could be successfully used." The authors further note:

> The focus on activity and incompleteness seemed to lead students to associate procedures more with the processes they describe rather than with the objects they are. Indeed, many students placed procedures in a category separate from other objects, viewing procedures as a different sort of thing— or, perhaps, not as a 'thing' at all.

Eisenberg and his colleagues suggest that the origins of this misperception of procedures may be related to linguistic structures (*e.g.* noun/verb distinctions) in English or to previous programming experience. DiBiase (1995), however, proposes that misconceptions about the *mathematical* concept of functions might be the true cause of the problem. Mathematical education research such as that described in (Harel & Dubinsky, 1992) may prove enlightening in this area.

Attitudes that Interfere with Learning

We move on to a somewhat different category of interference with learning, that of mistaken beliefs or attitudes. One example is *confirmation bias*, the tendency of people to test a hypothesis with data that confirms rather than refutes the hypothesis. This tendency is a serious handicap to thorough software testing. Leventhal, Teasley, & Rohlman (1994) found confirmation bias even in expert programmers. They showed, however, that a more complete problem specification yielded better testing, since the expert subjects primarily derived their tests from the specification. The presence of errors in the program being tested may also have an influence on test coverage, although results were mixed in the Leventhal experiments.

Thus one way to reduce confirmation bias is to produce more complete problem specifications. Two other suggestions from (Leventhal *et al.*, 1994) are better education about the value of negative tests, and standards that require each positive test to be "balanced" by a negative test.

Previous education can also bias students in unproductive directions. Hoadley, Linn, Mann, & Clancy (1996) explored the tendency of students to reuse previously studied or developed code. Beliefs about code reuse explained many of the failures to reuse observed in the study: students less inclined toward reuse cited the difficulty of comprehending code of others, their own poor memories, and a belief that code reuse was plagiarism. Hoadley suggested that instructors model the process of reuse, in particular reassuring students that code copying is "legal".

Fleury, in the most extensive investigation of this type (Fleury, 1993), conducted in-depth interviews of twenty-three students in introductory programming courses, along with four CS graduate students. Her key finding was that novices strove to *avoid* complexity, while experts aimed to *manage* complexity. Here are some examples of these differences:

> Some students expressed the belief that arrays and arrays of records made the programs containing them difficult for people to read. ...[Quote from student] S11: 'It looks like too much is happening, you know, when you first

read the program. Oh, no! Arrays! Records! Everything else!' … Experts, on the other hand, advocated the use of such data structures as arrays of records, and explained that they were expected as an alternative to freezing the program to work with a fixed number of individual variables.

Students see debugging as finding and fixing errors, with the emphasis on finding the errors. … Many students see debugging as essentially completed when they have found the procedure in which the (assumed single) error resides. … The experts viewed debugging and testing differently, … [explaining] that program comprehension is a necessary part of debugging and testing … [and describing] the practical importance of fixing all the bugs.

Many students see program maintenance, modification, and extension as academic exercises. They are concerned about minimizing typing and about facilitating superficial changes to program syntax. … Experts, on the other hand, assume that programs will be modified and extended over the course of the program lifetime in response to real-world events.

Fleury noted the importance of instruction and course activities in countering this bias. In particular, she suggested more exercises in *reading* code, to sensitize students to the benefits of techniques of complexity management.

Addressing Misconceptions

General approaches
As noted earlier, misconceptions spring from a learner's efforts to link knowledge. Remedies must focus on that linkage to foster more coherent understanding.

In particular, mere confrontation of a student's misconception with contradictory evidence is unlikely to succeed. Eylon & Linn (1988) note that students may ignore contradictions to their ideas or modify observations to defend their views. Smith *et al.* (1993) observe in addition that

Confrontation is also problematic as an instructional model. In contrast to more even-handed approaches to classroom discussions where students are encouraged to evaluate their conceptions relative to the complexity of the phenomena or problem, confrontation essentially denies the validity of students' ideas.

Better strategies encourage *knowledge integration*, the dynamic process of linking, connecting, distinguishing, organizing, and structuring ideas about a given concept. Techniques for doing this include bridging analogies, self-explanation, and group learning via discussion and experimentation. Linn (1995) proposes a framework for instructional design called "Scaffolded Knowledge Integration" that selects models that build on student intuitions, encourages knowledge integration, and helps students view themselves as continuously refining scientific ideas. Examples of how this might be done include making thinking visible through a variety of mechanisms (dynamic models, multiple representations, case studies, tutoring), encouraging autonomous learning, and providing social support for learning. Linn *et al.* (2003) also describe the use of *pivotal cases* to aid knowledge integration; these cases present important comparisons of scientific reasoning that enable students to use their inquiry skills to reorganize their ideas.

Slotta & Chi (2002) reported on another example of knowledge integration, intended to overcome physics misconceptions relating to *emergent processes*. Emergent processes were characterized as follows:

- They have no clear cause-and-effect explanation.
- They involve a system of interacting components seeking equilibrium among several constraints.
- In an emergent process, certain constraints behave as they do because they are actually the combined effect of many smaller processes occurring simultaneously and independently within the system.
- Emergent processes have no beginning or ending, even if they arrive at an equilibrium position.

Slotta and Chi administered a training module that highlighted the emergent process of diffusion. Subjects who worked through this module, then through an explanation of electricity, learned the latter material better than subjects given a more typical treatment of diffusion.

Approaches relating to computer programming

For misconceptions relating to specific computing agents, an obvious approach is to reveal more of the computing model. One must take care, however, to make clear the correspondence between the model presented and the modeled process. For example, the dynamic process of program execution is difficult to convey with a static description. Rajan (1990) listed some desiderata for a computing model:

- simplicity;
- consistency throughout the system of the presentation of information about program execution;
- transparency to the proper level of detail.

All these properties aid the knowledge integration process.

George (2000a, 2000b) created an environment named EROSI ("Explicit Representer Of Subprogram Invocations") that provides a diagrammatic trace of a Pascal program. When a subprogram is called, a new frame containing the subprogram's code is displayed. EROSI improved learning of the "copies" model of recursion that experts hold. However, when subjects who had used EROSI were forced to work without it to understand recursive code, many of them reverted to the incorrect "loop" model.

Stasko and colleagues at Georgia Tech have long studied the benefits of animations used as an aid to learning algorithms. Their results have been mixed, which suggests that an animated computing model would not necessarily help with related misconceptions. Kehoe, Stasko, & Taylor (2001), however, note that animations help *motivate* students, reducing the intimidation of a complex computing model.

Levy, Lapidot, & Paz (2001) have explored *preconceptions* of recursion as a way to study and possibly short-circuit the formation of misconceptions. High school students just being introduced to recursion engaged in a four-phase activity:

- exposure to recursive phenomena;

- group work to classify the phenomena;
- group presentation of its classifications, categories, and criteria;
- reflective class-wide discussion.

The discussions yielded rich and complex conceptual information, suggesting that social interaction can stimulate elaboration of concepts, and providing valuable clues about how students form their own models of recursion.

Ben-Ari (2001) describes a number of other examples of computing misconceptions and how they might be addressed. He also suggests some principles for applying the constructivist approach to computer science education in general.

Current Research Directions

Users of modern programming languages and environments have no shortage of opportunities to form misconceptions. Java/C++ constructs such as iterators, generic functions, templates, and inheritance all involve subtlety sufficient to confuse students, and unearthing the sources of confusion will continue to be a fruitful activity. Similarly, class libraries included in programming environments are a source of mystery. We have noticed students trying to hash the result returned by the hashCode method in Java's HashTable class—what's going on there?

Misconceptions in many of the "old" aspects of programming are still, to our knowledge, unexplored. For example, students commonly write code of the form

```
if (a > limit) return true
else return false;
```

It would be nice to know why they are avoiding the simpler shorter

```
return a > limit;
```

Research into recursion is still active, and can profitably be extended into study of how more complicated recursions are (mis)understood. Pointer bugs are the bane of students in a data structures course; a satisfactory explanation of how students comprehend pointers would be a big help to teachers.

Applications of distributed computing and concurrent processing have been covered in operating systems courses for years. They have recently begun to appear in lower-level courses, however. Common patterns of errors in mismanaging concurrent processes are just beginning to appear (*e.g.* in Choi and Lewis (2000); Kolikant, Ben-Ari, & Pollack (2000); Kolikant (2001)). It's hard to think concurrently, and research is needed into what models help. Resnick (1996) and Guzdial (1995) have both noticed a *centralized mindset* among programmers designing simulations, that is, a tendency to assume the existence of a controlling object with leader-centered communication, rather than a more decentralized model. What is the detailed nature of this mindset, and how does it arise?

Our own work addresses other aspects of detecting and countering misconceptions. We recently reorganized our introductory programming course (Clancy, Titterton, Ryan, Slotta, & Linn, 2003) to use an online learning environment. The pedagogical impact of the change was to increase the proportion of supervised lab activities and online collaborative discussion, and to decrease the "step size" between activities. As a result, we have a far more detailed picture of each student (misconceptions, coping strategies, etc.), drawn from the recording of a higher proportion of students' work, the small step size from activity to activity, and the new collaborative activities. We noticed students encountering a number of difficulties that had previously been invisible in the old course organization of lab/lecture/discussion, for instance, the confusion about the roles of empty words and sentences mentioned early in this presentation. The online environment, based on the WISE learning environment developed by Marcia Linn and her colleagues at Berkeley (Linn *et al.*, 2003), supports the principles of the Scaffolded Knowledge Integration framework mentioned earlier. Its tools allow us to design activities through which students may refine their understanding through experimentation and discussion. We are excited by the prospects, already demonstrated, of better student learning and better instructor understanding of the process by which students come to understand the complexity of programming.

Finally, we need research into the process of programming knowledge integration in general. What tools and activities help? What opportunities are provided by advanced technology? What are the characteristics of a good "pivotal case"? How can we measure deep understanding? (These and a number of other suggestions are detailed in National Research Council (2000).)

Summary

Historically, computing technology has been far ahead of computer science pedagogy. Each new programming paradigm—procedural, functional, declarative, and object-oriented programming—has its own computational model, usage idioms, and so on. Windows/mouse-based graphical user interfaces are an order of magnitude more complex than their text-based predecessors. New applications present mysterious execution models.

Not surprisingly, each new technological advance comes with its own set of misconceptions that educators must be aware of in order to help students learn the advance effectively. Here, we surveyed a body of research into programming misconceptions and their causes, and suggested some approaches for addressing these misconceptions. Particularly important are activities that aid the process of *integrating* and linking knowledge. We hope this survey will provide useful background for future researchers in this area.

References

Anderson, J., Pirolli, P. & Farrell, R. (1988). Learning to program recursive functions. In M. Chi, R. Glaser, & M. Farr (Eds.), *The Nature of Expertise*. Hillsdale, NJ: Lawrence Erlbaum Associates.

Bayman, P. & Mayer, R. (1983). A diagnosis of beginning programmers' misconceptions of BASIC programing statements. *Communications of the ACM, 26*(9), 677-679.

Ben-Ari, M. (2001). Constructivism in computer science education. *Journal of Computers in Mathematics and Science Teaching, 20*(1). 45-73.

Bonar, J. & Soloway, E. (1989). Preprogramming knowledge: A major source of misconceptions in novice programmers. In E. Soloway and J. Spohrer (Eds.), *Studying the Novice Programmer*. Hillsdale, NJ: Lawrence Erlbaum Associates.

Brown, J. & Burton, R. (1978). Diagnostic models for procedural bugs in basic mathematical skills. *Cognitive Science, 2*, 155-192.

Brown, J. & VanLehn, K. (1980). Repair theory: A generative theory of bugs in procedural skills. *Cognitive Science, 4*, 379-426.

Choi, S-E. & Lewis, E. (2000). A study of common pitfalls in simple multi-threaded orograms. *SIGCSE Bulletin, 32*(1), 325-329.

Clancy, M., Titterton, N., Ryan, C., Slotta, J., & Linn, M. (2003). New roles for students, instructors, and computers in a lab-based introductory programming course. *SIGCSE Bulletin, 35*(1), 132-136.

Davis, E., Linn, M., Mann, L., & Clancy, M. (1993). Mind your P's and Q's. In C. Cook, J. Scholtz, & J. Spohrer (Eds.), *Empirical Studies of Programmers: Fifth Workshop.* Norwood, NJ: Ablex.

DiBiase, J. (1995). Examining the difficulty with thinking of functions as data objects: Misconceptions of higher order functions. Technical report CU-CS-791-95. Boulder, CO: University of Colorado, Department of Computer Science.

Dicheva, D. & Close, J. (1996). Mental models of recursion. *Journal of Educational Computing Research, 14*(1), 1-23.

Du Boulay, B. (1989). Some difficulties of learning to program. In E. Soloway and J. Spohrer (Eds.), *Studying the Novice Programmer.* Hillsdale, NJ: Lawrence Erlbaum Associates.

Eisenberg, M., Resnick, M., & Turbak, F. (1987). Understanding procedures as objects. In G. Olson, S. Sheppard, & E. Soloway (Eds.), *Empirical Studies of Programmers: Second Workshop.* Norwood, NJ: Ablex.

Eylon, B. & Linn, M. (1988). Learning and instruction: An examination of four research perspectives in science education. In *Review of Educational Research, 58*(4), 251-301.

Fleury, A. (1991). Parameter passing: The rules the students construct. *SIGCSE Bulletin, 23*(1), 283-286.

Fleury, A. (1993). Student beliefs about Pascal programming. *Journal of Educational Computing Research, 9*(3), 355-371.

Fleury, A. (2000). Programming in Java: Student-constructed rules. *SIGCSE Bulletin, 32*(1), 197-201.

Fung, P., Brayshaw, M., Du Boulay, B., & Elsom-Cook, M. (1990). Towards a taxonomy of novices' misconceptions of the PROLOG interpreter. *Instructional Science, 19*(4/5), 311-336.

George, C. (2000). EROSI: Visualising recursion and discovering new errors. *SIGCSE Bulletin, 32*(1), 305-309.

George, C. (2000). Experiences with novices: The importance of graphical representations in supporting mental models. In A. Blackwell & E. Bilotta (Eds.), *Proceedings of the Twelfth Annual Workshop of the Psychology of Programming Interest Group,* 33-44.

Guzdial, M. (1995). Centralized mindset: A student problem with object-oriented programming. *SIGCSE Bulletin, 27*(1), 182-185.

Haberman, B. & Kolikant, Y. (2001). Activating "black boxes" instead of opening "zippers": A method of teaching novices basic CS concepts. *SIGCSE Bulletin, 33*(3), 41-44.

Halasz, F. & Moran, T. (1982). Analogy considered harmful. In *Proceedings of Human Factors in Computer Systems,* 383-386, Gaithersburg, Maryland.

Harel, G. & Dubinsky, E. (1992). *The concept of function: Aspects of epistemology and pedagogy.* Washington, DC: Mathematical Association of America.

Harvey, B., & Wright, M. (2000). *Simply Scheme* (second edition). Cambridge, MA: MIT Press.

Hennessy, S. (1990). Why bugs are not enough. In M. Elsom-Cook (Ed.), *Guided Discovery Tutoring: A Framework for ICAI Research.* London: Paul Chapman Publishing Ltd.

Hoadley, C., Linn, M., Mann, L. & Clancy, M. (1996). When, why, and how do novice programmers reuse code? In W. Gray &D. Boehm-Davis (Eds.), *Empirical Studies of Programmers: Sixth Workshop.* Norwood, NJ: Ablex.

Holland, S., Griffiths, R., & Woodman, M. (1997). Avoiding object misconceptions. *SIGCSE Bulletin, 29*(1), 131-134.

Hook, K., Taylor, J. & Du Boulay, B. (1990). Redo "try once and pass": the influence of complexity and graphical notation on novices' understanding of PROLOG. *Instructional Science, 19*(4/5), 337-360.

Kahney , H. (1983). What do novice programmers know about recursion? In *Proceedings of the SIGCHI Conference on Human Factors in Computer Systems*, 235-239, Boston, MA.

Kehoe, C., Stasko, J., & Taylor, A. (2001). Rethinking the evaluation of algorithm animations as learning aids: An observational study. *International Journal of Human-Computer Studies*, *54*(2), 717-749.

Kolikant, Y. (2001). Gardeners and cinema tickets: High school students' preconceptions of concurrency. *Computer Science Education*, *11*(3), 221-245.

Kolikant, Y., Ben-Ari, M., & Pollack, S. (2000). The anthropology of semaphores. *SIGCSE Bulletin*, *32*(3), 21-24.

Kurland, D. & Pea, R. (1989). Children's mental models of recursive Logo programs. In E. Soloway and J. Spohrer (Eds.), *Studying the Novice Programmer*. Hillsdale, NJ: Lawrence Erlbaum Associates.

Laurillard, D. (1990). Generative student models: The limits of diagnosis and remediation. In M. Elsom-Cook (Ed.), *Guided Discovery Tutoring: A Framework for ICAI Research*. London: Paul Chapman Publishing Ltd.

Leventhal, L., Teasley, B., & Rohlman, D. (1994). Analyses of factors related to positive test bias in software testing. *International Journal of Human-Computer Studies*, *41*(5), 717-749.

Levy, D., Lapidot, T., & Paz, T. (2001). "It's just like the whole picture, but smaller": Expressions of gradualism, self-similarity, and other pre-conceptions while classifying recursive phenomena. In G. Kadoda (Ed.), *Proceedings of the Thirteenth Annual Workshop of the Psychology of Programming Interest Group*, 249-262.

Linn, M. (1995). Designing computer learning environments for engineering and computer science: The Scaffolded Knowledge Integration framework. *Journal of Science Education and Technology*, *4*(2), 103-126.

Linn, M., Clark, D. & Slotta, J. (2003). WISE design for knowledge integration. *Science Education*, 87, 517-538.

National Research Council. (1997). *Science Teaching Reconsidered: A Handbook*. Washington, DC: National Academy Press.

National Research Council. (2000). *How People Learn* (expanded edition). Washington, DC: National Academy Press.

Pattis, R. (1981). *Karel the Robot*. New York: John Wiley & Sons.

Pea, R. (1986). Language-independent conceptual "bugs" in novice programming. *Journal of Educational Computing Research*, 2(1), 25-36.

Perkins, D. & Martin, F. (1986). Fragile knowledge and neglected strategies in novice programmers. In E. Soloway & S. Iyengar (Eds.), *Empirical Studies of Programmers*. Norwood, NJ: Ablex.

Pirolli, P. & Anderson, J. (1985). The acquisition of skill in the domain of programming recursion. *Canadian Journal of Psychology 39*, 240-272.

Putnam, R., Sleeman, D., Baxter, J. & Kuspa, L. (1989). A summary of misconceptions of high school BASIC programmers. In E. Soloway and J. Spohrer (Eds.), *Studying the Novice Programmer*. Hillsdale, NJ: Lawrence Erlbaum Associates.

Rajan, T. (1992). Principles for the design of dynamic tracing environments for novice programmers. In M. Eisenstadt, M. Keane, & T. Rajan (Eds.), *Novice Programming Environments: Explorations in Human-Computer Interaction and Artificial Intelligence*. Norwood, NJ: Lawrence Erlbaum Associates.

Resnick, M. (1996). Beyond the centralized mindset. *Journal of the Learning Sciences*, 5(1), 1-22.

Scherz, Z., Goldberg, D., & Fund, Z. (1990). Cognitive implications of learning PROLOG: mistakes and misconceptions. *Journal of Educational Computing Research*, 6(1), 89-110.

Segal, J. Ahmad, K. & Rogers, M. (1992). The role of systematic errors in developmental studies of programming language learners. *Journal of Educational Computing Research*, 8(2), 129-153.

Sleeman, D., Putnam, R., Baxter, J. & Kuspa, L. (1988). An introductory Pascal class: a case study of students' errors. In R. Mayer (Ed.), *Teaching and learning computer programming: Multiple research perspectives*. Norwood, NJ: Lawrence Erlbaum Associates.

Slotta, J. & Chi, M. (2002). Overcoming robust misconceptions through ontology training. Manuscript submitted for publication.

Smith, J., diSessa, A., & Roschelle, J. (1993). Misconceptions reconceived: A constructivist analysis of knowledge in transition. *Journal of the Learning Sciences*, *3*(2), 115-163.

Spohrer, J., & Soloway, E. (1986). Analyzing the high frequency bugs in novice programs. In E. Soloway & S. Iyengar (Eds.), *Empirical Studies of Programmers*. Norwood, NJ: Ablex.

Taylor, J. (1990). Analysing novices analysing PROLOG: What stories do novices tell themselves about PROLOG? *Instructional Science*, *19*(4/5), 283-309.

Van Someren, M. (1990). Understanding students' errors with PROLOG unification. *Instructional Science*, *19*(4/5), 361-376.

Van Someren, M. (1990). What's wrong? Understanding beginners' problems with PROLOG. *Instructional Science*, *19*(4/5), 257-282.

VanLehn, K. (1990). *Mind Bugs*. Cambridge, MA: MIT Press.

Notes

1. The term "misconception" derives from the mismatch between the student's underlying concept and "reality" (e.g. the actual operation of a computational model). We will use this term despite evidence—see Smith, diSessa, & Roschelle (1993)—that they are key components of learning, and perhaps ought not be viewed so negatively.

2

Critical Enquiry in Computer Science Education

Tony Clear

Introduction

Critical enquiry is a term for a school of somewhat controversial research methods. Although very rarely used in the computer science (CS) field, it is a rather more common research approach in other disciplines, especially in education and various fields of social science research. Yet in the computer science discipline, the core methods of research offer a rather limited repertoire for the inherently transdisciplinary endeavour of CS education research. To support this different range of research topics and goals it is necessary to extend the traditional repertoire by borrowing methods from other disciplines. *Critical enquiry* represents one alternative family of methods, which can be used to support different forms of enquiry in CS education research.

Is There Research Beyond The "Normal Science" Paradigm?

Critical enquiry can be thought of, not so much as a research method or group of research methods, but as a distinct research "paradigm" (Kuhn, 1962), with its own worldview and set of beliefs about the nature of knowledge and truth. As a paradigm it may be positioned within the three broadly recognised paradigms of research (Melrose, 1996, Orlikowski & Baroudi, 1991, Carspecken, 1996 p. 20, Clear, 2001c) outlined below:

- The *scientific/scientistic* (sometimes also called the *objective* or the *positivistic* approach). This research paradigm is usually typified by the making of formal hypotheses, and the use of quantitative methods to assess their validity. Conclusions are normally drawn negatively, and stated in terms that imply that the hypothesis has not (for now) been disproven, so it can be assumed to hold true. [This represents the primary "normal science" (Kuhn, 1962) research paradigm within which most computer *scientists* have been schooled.]
- The *interpretivistic* (sometimes called *subjective* approach). This research method often tries to understand complex phenomena which cannot readily be analysed in a fragmented way, such as social systems, societies or aspects of individual lives. Hypotheses may or may not be used, but the measurement techniques often elicit opinions and feelings and involve more qualitative measures, in an attempt to more broadly understand the whole phenomenon under study.
- The *critical* method, which has an explicitly *emancipatory* mission, with an interest in addressing issues to do with power imbalances and liberation from unwarranted forms of constraint - thus a concentration on a particular aspect of the *benefit to mankind* dimension of research, as opposed to simply *adding to our stock of knowledge*. This method while basically *interpretivistic*, often combines some *positivistic* approaches.

How Might Critical Enquiry Contribute To Computer Science Education Research?

Enquiry using the critical method enables a different set of research questions to be addressed. It is particularly useful when investigating issues in which a power imbalance is present, leading to marginalisation of those involved in the situation and where research methods based upon traditional scientific assumptions will normally not challenge but merely reinforce an unsatisfactory status quo.

Some of the critical questions in CS education research relate to: the paucity of women studying the discipline (Camp, 2002); understanding the barriers to study for minorities in the discipline (Barker et al., 2002, Rocco, 1998, Billings, 2003); how to teach the discipline in a manner that empowers and motivates the learners (Robinson, 1994, Smith, Mann et al., 2001); the application of information technology in a manner that transforms the learning experience (Clear, 2000); the rise of consumerism in computing education (Clear, 2002c); the barriers to sharing the experience and insights of seasoned IT practitioners in the academy (Clear, 1999c, Clear & Young, 2002); adjustment strategies for both computing educators and learners from other cultures in an increasingly internationalised learning context (Billings, 2003, Chamberlain & Hope, 2003).

These tend to be broader issues than simply applying a new teaching technique or an innovative technology in teaching a specific computing subject. While evaluation of such educational innovations can in themselves prove challenging (cf. Almstrum et al., 1996, Bain, 1999), selecting and applying suitable research techniques to explore the set of questions posed above requires a much broader armoury. Critical enquiry with its holistic approach, its focus on analysing imbalanced power structures, how they are reinforced and sustained, and its often activist and interventionist strategies to effect meaningful change, provides a set of tools for researchers seeking to branch out beyond those approaches that may have served them in the Computer Science discipline itself.

So What Do We Mean By Critical Enquiry?

The research strand known as *critical enquiry*, has developed from the critical social theory, of Jurgen Habermas (cf. Habermas, 1972, 1984, 1989 and Held, 1980) and his colleagues in the "Frankfurt School" of critical social theorists, cf. Held. (1980). For a reader new to critical social theory the language can prove a barrier, typically being written so densely that as Carspecken (1996, p. 4) comments: "This has made work in the critical tradition basically inaccessible to a large number of people". The aim of this chapter is to provide an accessible introduction to critical theory, and situate its relevance within CS education research.

The critical social theory of Habermas is based upon his "theory of cognitive interests." In this theory Habermas as a neo-marxist defines human beings from a basis of historical materialism, as labouring, toolmaking and language using animals, the basic activities through which humans produce and reproduce their species. These it is argued furnish man with an *a priori* set of interests. These three interests it is argued (cf. Held, 1980) are: 1) the *technical* interest associated with tool making, 2) the *practical* interest in creation of knowledge so that control of objectified processes and maintenance of communication can occur to support the technical interest; 3) the *emancipatory* interest, a reflective interest, which enables insight into the character of knowledge itself. This interest in reason, in the human capacity to be self reflective and self-determining, to act rationally generates knowledge, which enhances autonomy and responsibility and is hence an *emancipatory* interest.

These three interests are said to unfold in three media - work (instrumental action), interaction (language) and power (asymmetrical relationships of constraint and dependency), and give rise to the conditions for the possibility of three sciences, the empirical analytic, the historical-hermeneutic and the critical.

Carr & Kemmis (1983) represent these interests diagrammatically thus:

Table 1 The knowledge constituted interests of Habermas

Interest	Knowledge	Medium	Science
Technical	Instrumental (causal explanation)	Work	Empirical-analytic or natural sciences
Practical	Practical (understanding)	Language	Hermeneutic or 'interpretive' sciences
Emancipatory	Emancipatory (reflection)	Power	Critical sciences

From this foundation Habermas (1984) developed his 'theory of communicative competence' in which he posits the notion of all speech as oriented to a genuine rational consensus, the *ideal speech situation,* which is rarely realized (cf. Wilson, 1997). This ideal speech situation then becomes the ultimate criterion of the truth of a statement or the correctness of norms, creating an underpinning for critical theory grounded in the very structure of social action and language. The notion of an ideal form of discourse then can be used as a standard for a critique of distorted communication. "It is Habermas's contention that in every communicative situation in which a consensus is established under coercion or under other similar types of condition, we are likely to be confronting instances of systematically distorted communication." (Held, 1980)

What Does It Mean To Be A "Modernist"?

While Habermas provides the theoretical basis for most critical social theory, his work is not without its critics. Positioning his work in the modernist versus postmodernist debate, reveals some contradictions in critical theory itself. Taket & White (1993) describe the phenomenon of modernism as resting "on a belief in the capacity of humanity to perfect itself through rational thought. The modern is exemplified by the criteria of progress and reason." Computer science therefore can be considered as a 'modern' discipline.

Modernism has been further described in two modes, 'the systemic and the critical'. In the *systemic* mode knowledge and information are organising principles in effecting social control and directing change. The *critical* mode of modernism, by contrast works against this mechanistic process with a liberating rather than controlling purpose. This purpose is effected by working to liberate the so called 'lifeworld', a difficult but crucial concept which is explained below.

This central concept of a 'lifeworld' describes a certain integrity of views derived from "life experiences and beliefs which guide attitudes, behaviour and action. The three main forms of the lifeworld are culture, society and personality" (Myers & Young, 1997). These lifeworlds are said to be held together through 'systems' and 'steering media' .

As an example, a mental health systems study by Myers & Young (1997) depicts the lifeworlds of doctors, nurses, systems analysts, hospital managers and Health service administrators, each having their own distinct characteristics.

The concept of *systems* in a lifeworld is important, as identifiable spheres of action, with economic and administrative systems being primary. These in turn are guided "by lifeworld concerns and held together by the steering media of money and power" (Wilson, 1997). Habermas proposes that it is normal for the steering media to steer the societal systems in ways consistent with life world demands. But "it is possible for the steering media to 'get out of hand' and to steer the societal systems in ways which is at odds with lifeworld demands. This process is called the 'internal colonization of the lifeworld.'" (Myers & Young, 1997).

In the mental health study above the lifeworlds of managers and health professional differed greatly, with managers focusing on efficiency and monitoring, and health professionals focussing on care. The systems designers in implementing a system to support managers' wishes, were inherently enrolled in an agenda which would "colonize the lifeworlds" of the health professionals. Such a process inherently represents a distortion of communication, and the role of modernist critical theory is to uncover and address such distortions, by developing theoretical approaches to enable collective emancipation, by improving the lot of others.

What On Earth Is A Postmodernist Anyway?

By contrast the postmodern perspective "attacks all that modernity has 'engendered'; for example civilisation, industrialization, urbanization and technology. It challenges the values and objects of modernism such as individual responsibility, liberal democracy, ...rationality, quality, evaluative criteria and impersonal rules" (Taket & White, 1993). Distrusting modernity and seeing it as an oppressive rather than a liberating force, it does not favour any one credo over another such as Marxism, capitalism, humanism, or Christianity, and is opposed to the enlightenment ideal of progress of mankind through science, dismissing it as a form of thought control, a totalizing "grand narrative".

Reflecting these distinctions in part, two lines of emancipatory thought in critical theory can be discerned, the first based upon the work of Jurgen Habermas, the second based upon the work of Michel Foucault. Foucault's approach shares the profound scepticism of this postmodern world, and is more concerned with providing tools through which individuals can make visible the hidden ways in which they are constrained by power structures and develop personally empowering strategies in response. "Fundamentally, the issue is *human* emancipation or *self* emancipation" (Brocklesby & Cummings, 1996). This then, seems a very individualistic model of critical theory.

Foucault does however acknowledge the significant role of institutions in giving life structure and pattern through regular forms that are amenable to rational analysis. Thus there may not be a truthful 'grand narrative', but there may well be compelling local narratives to be analyzed and worked with. The focus then is on the individual thinker in the local context, applying specific local knowledge. By this focus on the individual as a critical thinker, Foucault undermines the role of the expert as therapist, as the solver of others' problems. Brocklesby & Cummings (1996) "Local' people possess the *reason*, but after years of being conditioned to privilege and defer to, the world of experts they lack the *resolution* and *courage* to employ their own reason. Critical theory should enable individuals to regain such courage".

So what precisely do these different views mean for a critical researcher? The Habermas inspired model addresses situations of collective disempowerment, by the expert researcher undertaking a theoretical critique, but it seems a little detached. The step from critique to action appears missing, and the role of political action is left to the subjects of the research to initiate. By contrast the Foucault inspired model has no actively democratising motivation. In opposition to the desire of Habermas to reduce power differentials, in fact, Foucault sees power as an active and positive force productive of social relations. The Foucault model is potentially "empowering" in the sense that each individual can choose to apply the tools generated from critique to a form of self-liberation in their local context. But it seems rather bleak, and in the absence of a value system, or some overall goal of improvement, perhaps even pointless. So the contrasts between the two theories are stark -- hopeless idealism on the one hand and bleak nihilism on the other.

How Does Critique Engender Action?

In a move from these positions towards pragmatic activism, the research approach known as *action research* strives to offer one set of solutions. The Australian educational action researchers Carr & Kemmis (1983) define action research as:

> a form of self-reflective inquiry undertaken by participants in social (including educational) situations in order to improve the rationality and justice of (a) their own social or educational practices, (b) their understanding of these practices, and (c) the situations in which the practices are carried out. It is most rationally empowering when undertaken by participants collaboratively, though it is often undertaken by individuals and sometimes in cooperation with 'outsiders'

Action research activity is said by Carr & Kemmis (1983) to have two essential aims, both to *improve* and to *involve*. The focus of this improvement lies in three key areas: improving a practice; improving the *understanding* of a practice by practitioners and improving the *situation* in which the practice takes place. Three kinds of action research are delineated: technical, practical and emancipatory (or critical), mapping to the three broad research "paradigms", outlined at the beginning of this chapter. Yet through these different methods legitimate forms of knowledge may be determined which reflect the perceptions and beliefs of the inquirer. As "Habermas…points out, knowledge and human interests are interwoven, as reflected in the choice of methods and the ends towards which such methods are put" (Susman & Evered, 1978).

Where Has Critical Inquiry Been Used?

For researchers with an interest in applying the techniques of critical enquiry within CS Education research, there is a large body of critical literature in the education field. For instance the work on critical ethnography by Carspecken (1996) is one of a series of books on critical social thought edited by Michael Apple, a noted educationist and critical researcher within the education field (cf. Apple, 1979, 1983, 1986, 1993). Rocco (1998) offers a good example of a study into the role of "privilege" in adult education. In the educational technology area useful resources are: the review of evaluation paradigms for instructional design by Reeves (1997); the *Educational Technology* Special Issue on The Ethical Position of Educational Technology in Society (Yeaman, 1994); and Clear (2002c) is an example of an Educational Technology/CS Education Research article applying a critical perspective. In the Operations Research Literature Taket (1993) & White (1994) explore the nature of expert power as employed by traditional operations researchers, and the distinction between the modern and postmodern perspectives. In the Information Systems literature there are a wide range of resources applying a critical social perspective (Hirscheim & Klein, 1989, Flood & Ulrich, 1991, the *DATABASE* Special Issues on Critical Analyses of ERP Systems (Howcroft and Truex, 2001, 2002), the Journal of Information Technology special issues on critical research in information systems (Brooke, 2002a, 2002b), the Information Technology and People special issue on gender and IS (Adam, 2002), and work by Myers (1995, 2000), Myers & Young (1997). In the IS World site Myers (2000) provides an online resource with comprehensive coverage of qualitative research methods which includes approaches to critical enquiry for scholars in Information Systems.

A long established critical strand within the computing literature is the work based upon the Scandinavian social democracy movement, and participatory design, cf. the *Communications of the ACM* special issue on participatory design (Kuhn & Muller, 1993). There is a strong critical thread within the action research literature in education, where work with groups of educators to change the status quo is a frequent emphasis of such research, cf. Carr & Kemmis (1983), Zuber-Skerrit (1996), Melrose (1996, 2001). Participatory action research with communities in such contexts as developing countries is explored by Elden & Chisholm (1993) in a special issue of *Human Relations*. In nursing research where the issues of institutional and professional power and patient needs often conflict, there is a well-developed body of research from a critical perspective (cf. Campbell & Bunting, 1991, Duffy, 1985, Kaminski, 2002, Browne, 2000, Mill et al. 2001). In the nursing discipline "creating a safe environment to talk about sexuality" (Glass & Walter, 1998) represents an example of one sensitive issue and a research approach by which it may be explored. Feminist literature too has its own strong social critical strand, exploring issues to do with gender, difference, societal structures and power (cf. King, 1994, Switala, 1999, Hedges, 1997). Topics such as "prostitution as work" (O'Neill, 1996), "cyberfeminism" (Gur_Ze'ev, 1999), and "feminist pedagogy" (Christie, 1997) are examples of critical feminist writing.

Why A Critical Approach?

Critical modes of enquiry have developed in a broad range of disciplines especially the social sciences, where the perceived 'objectivity' and 'neutrality' of traditional scientific modes of enquiry have been increasingly called into question. In fact, Carspecken (1996 p. 7) argues that "much of what has passed for neutral objective science is in fact not objective at all, but subtly biased in favour of privileged groups". He uses the history of intelligence tests as one example of the misuse of science by which minorities and the poor are frequently negatively labelled and by means of which "diagnosticians...unconsciously use the products of purportedly neutral research to support and expand a system that discriminates and oppresses."

For researchers in the CS education field it is vitally important to confront the inherent biases of an educational background strongly based, as it typically is, in traditional scientific beliefs and practices. In CS education we are dealing not simply with the issues of the discipline, but the nature of our students and their learning. This inherently involves the whole person, the cultural and institutional context, and the constraints imposed by contending forces within the learning situation. The CS education researcher needs an extended set of research approaches to enable inquiry into these broader issues of the social, the professional and the personal. Critical enquiry then, is a research orientation which offers a means for researchers to address issues to do with power, inequality, and forms of oppression, including those subtle forms of oppression sustained "by mainstream research practices" (Carspecken, 1996 p. 7).

For instance a critical perspective may have much to add to inquiries into inequity within the discipline, such as the research stream of "women in computing" (Camp, 2002, Cukier, Shortt et al., 2002). There are occasional examples of critical feminist approaches in this research endeavour (Estrin, 1996, Adam, 1996, Adam, 2002) but they are rare in a computer science context, perhaps because "professorial women in S&E fear that any commitment to feminist studies will make them appear peripheral to traditional science and lessen their chances for promotion and tenure" (Estrin, 1996). Yet critical enquiry enables a different stance to be adopted by the researcher, and offers a different set of tools and techniques to support broader forms of enquiry and critique.

Nonetheless critical enquiry in the service of an emancipatory research agenda must be entered into with some care, as it is itself open to criticism. Bishop for instance has commented, "Within the neo-Marxist emancipatory paradigm, a position developed to critique the 'distanced', 'objective' and impositional positivist paradigm, there is an inherent tendency for researchers themselves to initiate emancipatory research for those whom they consider to be oppressed and to direct attention to the possibilities for 'social transformation'. The intellectual arrogance of such theory-driven emancipationists has contributed to a new form of evangelism" (Bishop, 1996, p. 56).

Illustrative Cases

The three cases that follow present examples of critical enquiry in CS education, and briefly overview the nature of the enquiry, the basis and methods for its conduct, and the techniques applied. These may suggest ways in which a critical research approach could be adopted, by those interested in exploring further.

International Collaborative Learning – Learning as Transformation?

This case profiles the author's own research, conducted as an ongoing action research programme, involving international collaboration between students in New Zealand and Sweden, through a series of groupware trials in which the students work together in virtual groups to achieve common goals, (Clear, 1998, 1999a, 1999b, 2000, 2002a, 2002b) and (Clear & Daniels, 2000, 2001). The focus of this case is the action cycle of semester 2 1999, the second international trial between Auckland University of Technology (AUT) and Uppsala University which was reviewed in the author's thesis (Clear, 2000).

The dual cycle action research model of McKay and Marshall (1999, 2001), depicted in table 2 below, offers a useful framework within which to analyse the research. This model deliberately distinguishes between the real world practical problem solving elements of action research and its research oriented dimensions.

Table 2: Elements of Dual Cycle action research

Research Interest	Problem Solving Interest
A - a real world problem situation potentially of interest to the research themes of the researcher	P - a problem situation in which we are intervening
M_R - research method	M_{PS} - problem solving method
F - a theoretical framework	

In the research cycle reviewed here the distinct elements of the research are identified in table 3 below, applying the model of McKay & Marshall (1999, 2001). This summary of the research cycle, shows the relative complexity of the activity, the combined theoretical and problem solving dimensions of the research, and a degree of uncertainty regarding the research method itself. This partly reflects the fact that this framework is basically an analytical structure, overlaid upon the research after the event. It also results from a process of personal reflection by the author about the degree to which the research truly represented a *critical* or *emancipatory* model of action research. In a *practical* action research model, the researcher facilitates reflection by individual practitioners upon some aspect of their practice. In an *emancipatory* action research model a community of practitioners jointly negotiate goals and work to effect changes in the status quo. This issue will be explored further below.

Table 3: elements of the action research intervention Auckland - Uppsala second collaborative
      trial, Semester two 1999

Element	Description
F (Framework)	*(Theoretical bases informing conduct of the research)* Problem Based learning Adaptive Structuration Theory & Extended AST An Integrative Model of Group Interaction
M$_R$ (Research Method)	Practical Action Research, (Loosely framed), combining elements of Emancipatory action research
M$_{PS}$ (Problem solving method)	Practical Action Research, (Loosely framed), Prototyping
A - (problem situation of interest to the researcher)	To explore the structuring process for discussions and other communication, coordination, and collaboration facilities using the generic collaborative database To explore the moderator's role, the role of structure and the facilitation process using the collaborative database To improve understanding of groupware and Lotus Notes features, how to apply them, Notes' technical infrastructure and development process Alpha? Testing & Improving functionality in the prototype collaborative database To evaluate the effectiveness of the design concept of the database and explore the appropriation processes used by individuals and groups To explore methods of linking research and teaching
P - a problem situation in which we are intervening	Improving teaching & learning Developing student capabilities in teamwork, cross cultural communication and use of IT Providing an interesting & meaningful learning experience Using the collaborative database to teach and practically demonstrate key concepts of groupware and group decision support To perform a group ranking task Validating viability of collaborative databases for use by work teams or students engaged in international groupwork

During the "reflect" phase (Carr & Kemmis, 1983) of the action research cycle, or
the phase of "specifying learning" (Susman & Evered, 1978) it became apparent

to the author that there were some tensions inherent in the research context. These potentially invalidated the notion of the research as a model of true *emancipatory* action research. Now in order to support this process of *reflection* and *specifying learning* from the action cycle, as in any research model, the relevant data had to be selected and appropriate analysis had to be undertaken. But what is the nature of data in critical action research? In the process of exploring this question a taxonomy of data was derived, and these differing forms of data were represented in the structure of the author's thesis. "The early chapters presented the *historical and contextual* data, intermediate chapters the *process* of the action research, and the following chapters addressed the *empirical* and *evaluative* forms of data. This taxonomy is proposed in the table below, with illustrative examples given for each class of data." (Clear, 2000)

Table 4: A taxonomy of data in critical action research

Taxonomy of Data Types for Critical Action Research Projects			
Historical and Contextual	**Process**	**Empirical**	**Evaluative**
Examples of forms of data in this project			
Various AUT internal documents	Selected journal articles	Group membership details	Lecturer & course appraisals
Mission statements	Instructions & Timeline for Collaboration	Online evaluation questionnaires	Reflective reports, conference & journal articles
Research reports	Participant Information Sheet	Scoring, individual & group ranking entries	Student assignment reflective analyses
Strategic plans	Consent form	Online logbook entries	Reeves analysis in class
Teaching & Learning Development Plans	Complaint correspondence	In class email survey results	Personal reflection
Policy documents	Ethics approval documents & correspondence	Discussion postings & email messages	Reflective exam questions & Student responses
Programme reports	Database design notes & features	Attached files	Latent Discourses
Newspaper & magazine articles	Discussions in class and related email	Design proposals	Technical reports
Correspondence - research grants, innovative teaching awards etc.	Class presentations, module handbook, course handouts, course text extracts	Website links	Journal articles (online & offline)
newsletters	Database changes		Dilemmas
	Database entries		Emancipatory questions

In analysing the *evaluative* forms of data, an inherently difficult question is the issue of how and where to focus the analysis. Two useful techniques were dilemma analysis (McKernan, 1991) and critical incident technique (Chell, 1998). Dilemma analysis was used as a mechanism to tease out opposing poles of an issue, as a dialectic technique to identify significant tensions and explore the wider social structures of which they were part. The incidents isolated by critical incident technique, provided grounded data to inform the dilemma analysis.

The benefit of this method was its ability to draw the links between discrete, tangible events and broader societal structures, and thus bring into effect the principle of the *hermeneutic circle* (Klein & Myers, 1999). The *hermeneutic circle* is a key analytic principle in qualitative research, which provides for a form of triangulation of findings by confirming consistency of interpretations between the part & the whole. Hermeneutic analysis has been used heavily in researching biblical texts, and I like to think of it as the "zoom-in", "zoom-out" principle, whereby the researcher looks at the detail in depth and then zooms back out to the big picture to check for consistency of findings. The cycle may repeat several times until the meanings cohere. As an example of this form of analysis, during the research programme the author received a student complaint about the research project, its relevance to the course, and the fact that students were "customers" and "locked into the degree." This critical incident sparked considerable personal reflection.

One key issue that emerged was the power differential inherent in the teaching/student relationship. Consequently was this transformative model of learning a jointly chosen course of action, in which students and teachers became co-researchers using Information Technology to enable new forms of learning experience. Did it represent an emancipatory action research model, or merely a teacher imposed piece of whimsy, resented because it did not directly generate credit towards the course.

The other key issue arising from this incident was the dilemma represented in the form of two broader "discourses", the *discourse of enterprise* versus the *discourse of community* (Clear, 2002c). Foucault's (1980) concept of a *discourse* is described below.

> A discourse is a regulated system of statements and practices that defines social interaction. The rules that govern a discourse operate through language and social interaction to specify the boundaries of what can be said in a given context, and which actors within that discourse may legitimately speak or act" (Davies & Mitchell, 1994).

Thus a discourse both enables and constrains social action and acts to reinforce structures of power. But "a discourse is determined by community, it is also embedded in the larger framework of social relationships and social institutions" (Jennings & Graham, 1996). In this collaborative research project, situated within a wider social context, identifying some of the key discourses in operation, how they act to enable/restrain possibilities, and how they conflict with one another, has been a means of broadening our perspectives on the research undertaking.

The distinction drawn here between the *discourse of enterprise* versus the *discourse of community* is that between education as a personal economic good, an investment in the self, with educators as providers of services to student *customers*;

or education as a social responsibility, from which the community derives benefit and with many stakeholders interests to be balanced in the moral choices of professional educators. In AUT's model of quality "Education is a participative process, students are not products, consumers or customers. They are participants" (Horsburgh, 1996). Thus power does not lie solely with the student, the curriculum is not totally negotiable, so the emancipatory ability of educational action research may be limited, when the 'peer community' is not a community of equals, as it may be when working with professional colleagues. However by adopting an open approach, which makes visible these power imbalances, and by the use of a contracted learning model conducted within "a mutually acceptable ethical framework" (Susman & Evered, 1978), it is possible to conduct teaching, learning & research in a mode that has an emancipatory dimension.

Crashing a Bus Full of Empowered Software Engineering Students?

This case provides another New Zealand example, in which the teaching of a software engineering course on Otago Polytechnic's Bachelor of Information Technology (Smith et al., 2001), was informed by the attitudes and practices of an "empowering education" model (Robinson, 1994). The table below depicts a few elements of the model.

Table 4. Attitudes and Practices of Empowering Education

a) The teacher and students both teach and are taught by each other

b) The teacher is aware of not knowing everything and is open to the student's knowledge and experience which are actively valued

h) The teacher and the students together decide on programme content and revise and change it as their interests and needs change

j) The teacher and students form a collective Subject of the learning process, sharing joint ownership of the classroom life

The design of the software engineering course involved the application of a catastrophic change in project to an otherwise successful course. "The class was 'run over by a bus' and groups were required to swap projects halfway through development" (Smith et al., 2001). This design was intended to emulate the commercial environment in which individual software engineers would rarely see a project through its entire development, from planning to implementation.

The aim of this research was firstly to attempt to replicate and document in more detail the positive findings of an earlier study (Surendran and Young, 2001) with regard to swapping projects; and secondly to assess this practice in terms of Robinson's empowering paradigm.

In order to establish the initiative on a sound footing, ethics committee approval was sought and gained, both for the study itself using a control group (only six of the ten groups were required to swap); and for the need to change an assessment in the middle of the course - i.e after the run over by a bus event (ROBAB), about which students were only warned to the extent that their projects might take 'unpredictable turns'. While a degree of deception was inherent in this learning design, consistent with the concept of informed consent, students included the following statement in their management document, "All members of this group are aware that our experiences in undergoing this research project may be used in research into teaching methods for software engineering. We understand that identities will be confidential and that taking part in this research is entirely voluntary and will not affect in any way how we are treated by the lecturers in this course".

The evaluation of the project involved gathering information from a variety of sources, course evaluations, student results, student reviews of the process including the bus incident and self assessments of their projects against a predefined marking schedule, resulting in a combination of quantitative and qualitative data analysis. As the authors noted, with only ten groups participating in the study, a detailed statistical evaluation of the results was not possible.   Nonetheless student performance on the course and their feedback regarding the course compared favourably with results from the prior year's iteration (without ROBAB), with the majority of students in favour of repeating the exercise and 50% of them adding the condition that next year the ROBAB should hit all groups. Based upon this analysis and student comments from the other information sources, the first goal of the project was achieved - namely to replicate the Surendran & Young (2001) study about the positive effects of swapping projects.

The second goal (degree to which this was consistent with the empowering paradigm) required a different form of analysis, and adoption of a more critical reflective style. Results were analysed in terms of initial outcomes, and mid project responses compared to final responses. The initial outcomes demonstrated variously anger, enthusiasm and confusion amongst the students and some concerns about the unfairness of only some of the groups being hit by the bus. The unaffected groups were grateful and expressed relief.   During the mid-project logical design phase, groups had some negative responses to the workload of making sense of the foreign group's documentation, and some positive responses based upon the learning gained from seeing another group's material. They also struggled with their attachment to their own design and becoming motivated about the new material they had inherited. Even the unaffected groups reported some loss of enthusiasm. In the final responses students' opinions of the bus crash incident mellowed, and most came to see the value of documentation throughout the development process. A general finding appeared to be that the change had overall little impact on the final outcome of the project, with marks for the course following a relatively traditional pattern.

What was the impact of the bus crash on the course's conformity with the empowering paradigm? While not specifically identified in the paper as a methodological framework, a form of dialectic analysis was applied (i.e. A current situation or *thesis* is compared with an opposing or *antithetical* situation, and as the contradictions are resolved a *synthesis* is derived representing the result of the analytical process). A more formal model of this process can be found in Myers (1995) under the framework of *dialectical hermeneutics*.

In this case the issue was approached by framing and reflecting upon a critical question: does the control imposed by the lecturer and consequent loss of ownership of the project by the students outweigh the benefits of swapping? At first glance the tenets of table 4 above have been violated, with ownership of the process and choices reverting to the lecturer, and students being cast as victims of imposed circumstances. Students certainly voiced concerns over how they would be assessed and whether they would lose marks as a result of this perceived disruption in their learning. However the empowerment model does not preclude challenge, and expects that students will be actively engaged in meaningful teacher facilitated experiences.

Yet in creating a challenging situation, while the teacher may be empowering the students to achieve, students do not always immediately like things that threaten their passivity, so this may occasion some discomfort. It has been argued that personal control is a prerequisite for empowerment (Harris, 1994), and in this experience ROBAB students initially perceived a loss of self-control, but soon realized they could take charge of the situation, did have scope to exercise control over the remainder of their project, and gained enjoyment from the process. So it was argued that while the locus of control dislocates for a time it quickly returns and the overall feeling towards the course is positive.

By contrast the students who did not swap had their ownership removed by the threat of swapping but did not come to realize the benefits. Thus the adoption of this approach was seen to have imposed a phase of discomfort on the ROBAB groups, through which challenge they had emerged with an overall sense of achievement and a positive experience. By contrast those who had not had the benefit of the ROBAB challenge while initially relieved at being left alone, ended up overall less satisfied and less empowered from the experience.

Thus it can be seen that applying critical enquiry as a research approach into the effectiveness of an intervention in Computing Education involves a wide ranging form of analysis, involving the roles, actions and beliefs of the participants, the specific forms of data supporting that enquiry, and the linkages between the institutional and social forces that may constrain or prescribe the activities of the actors. Critical enquiry is of necessity holistic in its nature, and the tests for rigour in enquiry differ from those accepted in the classical science tradition.

Women Taking Positions within Computer Science?

In this case from the U.K., using feminist critiques of science, Stepulevage & Plumeridge (1998) analyse aspects of a computer science curriculum in an English "new" University, to show how in this context "Computer Science remains firmly situated within the domain of masculinist modern Western science" (ibid.). This study rejects what the authors consider the typical "gendered constructions" of much research into women in science, based upon a universality regarding women's positioning and the dichotomy between the hard logic-based approach and a soft context based one. This dichotomy which associates the *concrete and contextual* with women and the *abstract and logical* with men, is regarded by such researchers as contributing to women's exclusion from scientific domains.

In this study the authors deal more directly with how issues of power inform the positioning of women and men in CS education. Refuting the notion of their *identity as women* being the key problematic for the success of women in computing, the study argues that their *standpoint as women* provides a more illuminating framework for analysis. This concept of *standpoint* derives from the work of Collins (1991) in her work on black feminist thought and situated knowledges. *Standpoint* for this study assumes that women may have a common experience of subjugation, but each brings her own perspective to situations, so that self, community and society are seen through a personally shaped lens. "However common experience of oppression 'in no way guarantees that such a consciousness will develop among all women or that it will be articulated as such by a group…'" (Stepulevage & Plumeridge, 1998).

In this study the women participants were self-defining members of a minority (female computing students, representing some 24% of the computing student body), gaining entry to a body of knowledge in computer science. The study explores how "through their common experience of systematic exclusion from the enterprise of creating science and of subjugation as women that these women's standpoints can be identified". It is argued that this experience of exclusion or subjugation "enables them to engage with the rules in various ways, the tutor by attempting to integrate practice and abstraction, the students by developing alternatives to the given rules" (Ibid.).

In their analysis the authors investigate the *social construction* of a data structures course, to highlight the mutual construction of gender and computing within it. The method aimed to "unpack" this construction through a study of the course documentation, information on student backgrounds and outcomes, in-depth interviews with staff and students, and observations in both lectures and seminars. Thus the research method is a form of "deconstruction", an analytical technique originating in literary criticism with writers such as Jacques Derrida (1973) the French postmodern theorist. For Beath & Orlikowski (1994), "deconstruction of a document reveals the dependence of that text upon taken-for-granted assumptions that may suppress, distort, marginalize or exclude certain ways of thinking". A review of how to apply the techniques of deconstruction in research, can be found in the study by Beath & Orlikowski (1994) in which they deconstruct the *user - developer relationship* in information systems development.

In this analysis of women in computer science, the authors review feminist critiques of western science whereby science itself has been seen as a "masculinist rational practice", in which "the discourse of what is referred to as modern science remains firmly rooted in claims of ideological purity, neutrality and universality" (Stepulevage & Plumeridge, 1998). In this context then, women in computer science are inherently "outsiders" operating within a masculine domain. Adopting Collins' (1991) conceptualization of the 'outsider/within', these women students are then "outsiders in the enterprise of knowledge creation".

In a deconstruction of the epistemology of science by Harding (1986) three 'dogmas of science' are exposed, as a useful tool for the analysis of computer science. She cites these as: "science as sacred; physics as the paradigm of science'; and 'pure mathematics' as value-free".

In the first dogma science is seen as a story of creation that does not need to justify itself, and seen to act as a god's eye view rather than acknowledge that it works from a specific location, in a form of 'god-trick' producing, appropriating and ordering all difference. Science defends this position by asserting its separateness from society, with scientific facts distinct from social values.

In the second dogma with physics as the paradigm of science, the value-neutrality of physics is explored. In physics "the subject matter studied has been reduced to a simpler form and isolated from social constraints, i.e. the problematics of the everyday world, through the process of abstraction. The concepts and hypotheses of physics therefore, deny a need for social interpretation and explanations…There is no material context in which to frame a WHY question" (Stepulevage & Plumeridge, 1998).

The third dogma of science, wherein mathematical expressions are value-free, is shown to generate a form of purported neutrality through the reductionism necessitated by the experimental method, and the mathematical 'purity" derived from separating abstraction from reality, while failing to notice the problematic of the difficulty of reintegrating what is experimented upon back into its more complex social source. This separation of the pure from the applied is said to allow the pure the privilege of 'god-tricks', whereby the concrete products resulting from the discovery and development of algorithms can exist outside the domain of computer science and there is no need for critical self-reflection, a process argued to be missing from scientific education.

Armed with this set of tools the authors then proceed to analyse the construction of computer science in the experiences of the tutors and students in a data structures course. The four images relating to computer science in the University prospectus are analysed, and it is noted that the text used to represent computer science students visually establishes "white' and 'male' as dominant. The photographs "forecast the potential enjoyment of abstract thought by white men, and the access to computers gained by white women and black men" (ibid.). Black women are absent, a construction consistent with the exclusion of black women from science. The representation of the one white woman, with her personal story, helps construct the 'woman student' as exceptional, someone very determined and willing to travel to get where she wanted.

In the interviews with students and tutors on the course, the logical nature of the thinking required in computer science was commented upon. Two men students observed that 'you have to be fairly logical minded' while a woman student observed that "you have to familiarize yourself with the way tutors think, how they operate'. Similarly in discussing why there were fewer women students, the tutors mentioned computer science being seen as male dominated because it was perceived as technical or a science, and the women observed that up until very recently computing was not an area for women at all, because it was safe to conform to the norm of what has gone before, and also noted that women had to work harder to prove themselves more.

The authors' analysis of this difference, exposes the men as insiders aware of the rules, the women as outsiders needing to learn them. The study proceeds to explore the style of programming, by analysing the tutors beliefs, with the male tutor's description situating programming within a paradigm of rule following and using a proven method. The gendering of programming is argued as evident in that the context-fitting aspects of developing a program for direct use by people, e.g. acceptance testing, documentation and consideration of the software lifecycle are not considered relevant to this programming unit. A distinction was drawn between the male tutor's approach to teaching programming through abstraction as consistent with a scientific discourse, and the female tutor's standpoint, which emphasized the building of confidence through practice and understanding of the rules, a stance more consistent with a feminist epistemology, valuing knowledge gained through experience.

These insights have been developed based upon questioning not what is it about women and women's lives that have kept them from doing science, but what is it about science? The use of critical enquiry as a research perspective has enabled a different set of questions to be asked, a different form of analysis to be conducted, and as a consequence a new and more wide ranging set of insights to be generated into a set of critical issues in CS education.

Critical Enquiry In Computer Science Education

As noted above, critical enquiry is a relatively unusual research approach within CS education, largely because it adopts a rather different value position, and the above cases represent a few of the known studies. Other relevant work is summarized briefly below, but this collection should be viewed as representative of a rather dispersed literature rather than exhaustive. As an indicator of the paucity of research using this paradigm in CS education, a journal search of the ISI Web Of Science citation indices (ISI, 2002) returned 369 citations for "critical theory", 32 for "critical theory and education" and no citations for "critical theory and computer science". A slightly better ratio resulted from a search of the ACM digital library, returning 56 citations for "critical theory" and 17 for "critical theory and computer science", but of the latter few have any educational focus, some would be considered writing within the Information Systems domain, and some apply a form of literary criticism in their argument. A search of the ERIC (ERIC, 2002) educational database likewise returned 165 entries for "critical theory and education", and no entries for "critical theory and computer science".

In spite of this paucity some articles in addition to the cases profiled here can be identified, but ranging across a diversity of countries and discipline sources. Submissions based upon this paradigm tend to find greater acceptance in education, education technology or information systems outlets. A few examples of relevant work are briefly profiled below, including selected publications from the author's own work.

Selected examples of the use of critical enquiry in CS education

A first group of writings by the author (Clear, 2001a, 2001b, 2002b, 2002c) explore the issues of power imbalances in the teacher/student relationship, the increase of consumerism in the tertiary education context, the consequential impact on tertiary education and the role of public higher education.   The role of Information Technology in education is investigated, as a positive force for transforming the learning experience, reducing power differentials, or at least making them explicit through joint enquiry; or as a negative force furthering these consumerist tendencies, and encouraging passive and receptive student roles.

In a second group of writings (Clear, 1999d, 2003, Clear & Young, 2001, 2002) the authors investigate the research process itself, in the New Zealand higher education context, with particular emphasis upon computing in the Polytechnic and 'new' University sectors.   They explore the construction of research itself using a deconstructive form of discourse analysis investigating power/knowledge in the research process.   They investigate the barriers to new researchers, and critique changes in Government policy, which threaten to undermine the funding base supporting the linkages between research, teaching and practice in the computing discipline.   The beliefs of new computing researchers in this sector are explored. The research is conducted through a critical and practical action research programme aiming to increase the research capabilities of new degree teachers (often from an IT practitioner rather than a research background).   The research itself has been used as a means of modeling a critical research paradigm within a series of workshops conducted for new researchers within the sector.

In a third study Mann & Buissink Smith (2001) applied Robinson's (1994) empowering model of education to four undergraduate classes on the Bachelor of Information Technology at Otago Polytechnic, software engineering, databases, information systems management and the capstone project.   Quantitative and qualitative forms of data were analysed, and the results shown to compare favourably with the characteristics of Robinson's empowerment model of education, suggesting that this mode of pedagogy had indeed been achieved in the courses.

In a UK Based study Dawson & Newman (2002) argue "empowerment is at least as important as knowledge acquisition and that IT is an ideal vehicle to empower people studying a variety of subjects at different educational levels".   Given the volatility of the IT discipline they argue that the most useful attribute they can give their students is the confidence to find their own solutions to a given IT problem, and to cope with the unexpected in an IT context.   While this notion of empowerment is not specifically grounded in, or referenced to, critical theory, a model of teaching and learning involving empowerment of students does at least demonstrate consistent aims.   They argue that an experiential learning strategy, based upon project work supports an empowerment strategy.   Four case studies are reviewed, a software engineering style undergraduate course, a workplace learning context for new graduates, a systems engineering sequence in a degree programme, and a high school programme for disruptive children using IT as an intervention to improve behaviour and learning outcomes. The authors argue the success of their empowerment approach, which they consider to be based upon an interpretivist philosophy, rather than a positivist one. While this categorization omits the critical evaluative paradigm, the goal of teaching the students "to learn how to learn" is nonetheless consistent with an emancipatory or critical philosophy.

In another work Alexander from an American University (2002) applies Habermas' (1984) theory of communicative action to a study in which student teams in a first year information systems cohort of 1600 were offered three different options for completing their work, including working as classroom based teams, face-to-face independent groups or virtual teams collaborating via email. Using an action research method, questionnaires and observation, including recorded face-to-face conversations, were used to collect data, which were then analysed against the forms of communicative action proposed in the Habermas model. Findings of the study were mixed, with varying levels of communication arising, a surprising lack of interest in adopting the virtual group mode of study, relatively high non participation rates in the virtual groups, clear issues related to students' maturity and as a residential university, limited student need to study in this mode, but some degree of success in maintaining a permanent record of group activity.

In a US based ethnographic study Barker et al., (2002) immersed themselves in the learning environment of two different IT programmes, one a traditional computer science major and the other a technology arts & media programme. In the process of observing ten courses over a one year period the researchers compiled 648 pages of fieldnotes recording: number of students attending, sex and appearance, physical layout of classrooms and seating arrangements; and descriptions of interactions (student-student and student –instructor) and those interacting (male/female and major). The researchers then applied content analysis to categorize the data into patterns and themes. From this analysis in relation to the computer science courses, two categories arose which were: an impersonal environment and guarded behaviour; and an informal student hierarchy. These factors contributed to the creation of a defensive social climate in which the impersonal classroom climate communicated rejection rather than acceptance of students and the informal student hierarchy stemmed from communication emphasising superiority rather than equality, generating a competitive learning environment in which students are at risk of criticism, rather than a safer environment in which students can make mistakes and learn.

Conclusion

In conclusion it can be seen that critical enquiry is a research orientation, supported by a wide repertoire of methods, which can be used in diverse contexts relevant to CS education research. A set of new, holistic and frequently challenging questions can be addressed using critical enquiry. A deeper form of enquiry into power relations within the learning environment, and into the innate learning culture within a computer science context can be achieved by conducting research in a mode of critical enquiry. The findings from such research may spur us to seriously rethink the way in which the discipline is taught. This is an inherently challenging task, requiring considerable self-critical reflection, given the implicit and deeply rooted nature of beliefs in the computer science discipline and the associated teaching and learning cultures. Then of course, having decided that a change were warranted, the process of achieving that across the whole CS education teaching community would require another approach altogether. A critically informed community development and change framework such as critical action research might provide such a research model. But the inherent challenges of critical research present themselves when we consider how we might really engage the CS education community to effect meaningful change.

References

Adam, A. (1996). Constructions of Gender in the History of Artificial Intelligence. *IEEE Annals of the History of Computing, 18*(3), 47-53.

Adam, A. (2002). Guest Editorial: Special Issue on Gender and IS. *Information Technology and People, 15*(2), 1-4.

Alexander, P. (2002). *Teamwork, Time, Trust and Information.* Paper presented at the 2002 Annual Research Conference of the South African Institute of Computer Scientists and information Technologists on Enablement Through Information Technology, Port Elizabeth.

Almstrum, V., Dale, N., Berglund, A., Granger, M., Little, J. C., Miller, D., et al. (1996). *Evaluation: turning technology from toy to tool. Report of the working group on Evaluation.* Paper presented at the Integrating Technology into Computer Science Education Conference, Barcelona, Spain.

Apple, M. (1979). *Ideology and Curriculum.* London: Routledge and Kegan Paul.

Apple, M. (1983). *Education and Power.* London: Routledge and Kegan Paul.

Apple, M. (1986). *Teachers and Texts: A Political Economy of Class and Gender Relations in Education.* New York and London: Routledge.

Apple, M. (1993). *Official Knowledge: Democratic Education in a Conservative Age.* New York and London: Routledge and Kegan Paul.

Bain, J. (1999). Introduction (to the special Issue on Evaluation). *Higher Education Research & Development, 18*(2), 165-172.

Barker, L., Garvin-Doxas, R., & Jackson, M. (2002, March). *Defensive Climate in the Computer Science Classroom.* Paper presented at the Thirty Third SIGCSE Technical Symposium on Computer Science Education, Northern Kentucky.

Beath, C., & Orlikowski, W. (1994). The Contradictory Structure of Systems Development Methodologies: Deconstructing the IS-User Relationship in Information Engineering. *Information Systems Research, 5*(4), 350-377.

Billings, D. (2003). Did they Fail or were they Pushed? Student Retention and Success Initiatives in Tertiary Education. *NZ Journal of Applied Computing & IT, 7*(1), 17 - 22.

Bishop, R. (1996). *Collaborative Research Stories; Whakawhanaunatanga.* Palmerston North: Dunmore Press.

Brocklesby, J., & Cummings, S. (1996). Foucault plays Habermas: an alternative underpinning for critical systems thinking. *Journal of the Operational Research Society, 47*(6), 741- 754.

Brooke, C. (2002a). Editorial: Critical research in information systems: issue 1. *Journal of Information Technology, 17*(1), 45-47.

Brooke, C. (2002b). Editorial: Critical research in information systems: issue 2. *Journal of Information Technology, 17*(4), 179 -183.

Browne, A. (2000). The Potential Contributions of Critical Social Theory to Nursing Science. *Canadian Journal of Nursing Research, 32*(2), 35-55.

Camp, T. (Ed.). (2002). *Special Issue Women and Computing* (Vol. 34). New York: ACM.

Campbell, J., & Bunting, S. (1991). Voices and paradigms: Perspectives on critical and feminist theory in nursing. *Advances in Nursing Science, 13*(3), 1-15.

Carr, W., & Kemmis, S. (1983). *Becoming Critical: Knowing Through Action Research.* Melbourne: Deakin University press.

Carspecken, P. (1996). *Critical Ethnography in Educational Research*. New York: Routledge.

Chamberlain, B., & Hope, B. (2003). *Integrating international students into computing classes: issues and strategies*. Paper presented at the 16th Annual NACCQ Conference, Palmerston North, New Zealand.

Chell, E. (1998). Critical Incident Technique. In G. Symon & C. Cassell (Eds.), *Qualitative Methods and Analysis in Organisational Research*. London: Sage.

Christie, A. (1997). Using E-mail Within a Classroom Based On Feminist Pedagogy. *Journal of Research on Computing in Education, 30*(2), 146-176.

Clear, T. (1998, 17-21 August). *A Generic Collaborative Database - Part of a strategy to internationalise the curriculum and develop teamwork and communication capabilities*. Paper presented at the *The 3rd Annual Conference On Integrating Technology Into Computer Science - ITICSE'98*, Dublin City University Ireland.

Clear, T. (1999a, Jun 27 - Jul 1). *A Collaborative Learning Trial between New Zealand and Sweden - Using Lotus Notes Domino in Teaching the Concepts of Human Computer Interaction*. Paper presented at the The 4th Annual Conference On Innnovation & Technology In Computer Science Education, Cracow Poland.

Clear, T. (1999b, Jun 19-24). *International Collaborative Learning - The Facilitation Process*. Paper presented at the ED-MEDIA '99 - World Conference on Educational Multimedia, Hypermedia and Telecommunications, Seattle, Washington.

Clear, T. (1999c). "Eating our seed corn" or restricting crop diversity? *SIGCSE Bulletin, 31,* 15-16.

Clear, T. (1999d). A Critical Perspective On Research In New Zealand Polytechnics - Polysyllaby In The Polytechnic Sector. In G. Muriwai (Ed.), *Business Performance Research And Development Centre -Working Paper Series* (Vol. 1;1, pp. 1-22). Auckland: Faculty of Business, Auckland University of Technology.

Clear, T. (2000). *Developing and Implementing a Groupware Application to Support International Collaborative Learning*. Unpublished M. Phil., Auckland University, Auckland.

Clear, T. (2001a, June 2001). E-Learning to Whose Agenda? The Discourse of Enterprise vs. the Discourse of Community. *OLS News, 39-40.*

Clear, T. (2001b). Using Web Based Groupware for Active Student Feedback in the Learning Environment. *NZ Journal of Applied Computing & IT, 5*(1), 14 - 18.

Clear, T. (2001c). Research Paradigms and the Nature of Meaning and Truth. *SIGCSE Bulletin, 33,* 9-10.

Clear, T. (2002a). *Avatars in Cyberspace - A Java 3D Application to Support Performance of Virtual Groups*. Paper presented at the Innovation and Technology in Computer Science Education, Aarhus, Denmark.

Clear, T. (2002b). E-Learning or e-Lemmings? Who pipes the tune? *CSS Journal (Computers in the Social Studies), 10*(2), 1-7.

Clear, T. (2002c). E-Learning: A Vehicle for E-Transformation or Trojan Horse for Enterprise? Revisiting the Role of Public Higher Education Institutions. *International Journal on E-Learning, 1*(4), 15-21.

Clear, T. (2003). TEAC Research Funding Proposals Considered Harmful: ICT Research at Risk. *NZ Journal of Applied Computing and IT, 7,* 23-28.

Clear, T., & Daniels, M. (2000, Oct 18-21). *Using Groupware For International Collaborative Learning*. Paper presented at the The 30th American Society for Engineering Education/Institute of Electrical and Electronics Engineers Frontiers in Education Conference, Kansas, Missouri.

Clear, T., & Daniels, M. (2001, June 25 - June 27). *A Cybericebreaker for an Effective Virtual Group?* Paper presented at the The 6th Annual Conference On Innovation and Technology In Computer Science Education (ITiCSE), University of Canterbury, Kent.

Clear, T., & Young, A. (2001, 26-27 November). *Developing Novice Researchers' Understandings of Research*. Paper presented at the Innovations and Links: Research Management and Development & Postgraduate Education Conference. [Online]. Available:
http://www.aut.ac.nz/conferences/innovation/papersthemeone/clearpaperone.pdf [3 October 2002]. Auckland University of Technology, Auckland.

Clear, T., & Young, A. (2002). Met a Researcher? Research Paradigms Among Those New to Research. *NZ Journal of Applied Computing and IT, 6*(1), 18-25.

Collins, P. (1991). *Black Feminist Thought: knowledge, consciousness, and the politics of empowerment*. New York: Routledge, Chapman & Hall Inc.

Cukier, W., Shortt, D., & Devine, I. (2002). Gender and Information Technology: Implications of Definitions. *SIGCSE Bulletin, 34*(2), 142 -148.

Davies, L., & Mitchell, G. (1994). The Dual Nature of the Impact of IT on Organizational Transformations. In R. Baskerville, S. Smithson, O. Ngwengyama & J. DeGross (Eds.), *Transforming Organisations with Information Technology*. North Holland: Elsevier Science IFIP.

Dawson, R., & Newman, I. (2002). Empowerment in IT Education. *Journal of Information Technology Education, 1*(2), 125 - 141.

Derrida, J. (1973). *Speech and Phenomenon*. Evanston, Illinois: Northwestern University Press.

Duffy, M. (1985). A critique of research: a feminist perspective. *Health Care for Women International, 6*, 341-352.

Elden, M., & Chisholm, R. (1993). Emerging Varieties of Action Research: Introduction to the Special Issue. *Human Relations, 46*(2), 121-142.

ERIC. (2002). *ERIC Database*. Retrieved Nov 5, 2002, from http://www.askeric.org/Eric/

Estrin, T. (1996). Women's Studies and Computer Science: Their Intersection. *IEEE Annals of the History of Computing, 18*(3), 43-46.

Flood, R., & Ulrich, W. (1991). Conversations on Critical Systems Thinking. In R. Flood & M. Jackson (Eds.), *Critical Systems Thinking*. Chichester: John Wiley.

Foucault, M. (Ed.). (1980). *Power/Knowledge Selected Interviews and Other Writings 1972 - 1977*. New York: Pantheon.

Glass, N., & Walter, R. (1998). Creating a safe environment to talk about sexuality: Nursing educational research as the empowering strategy. *The Australian Electronic Journal of Nursing Education, 3*(2), 1-15.

Gur-Ze'ev, I. (1999). *Feminism, Education and Critical Theory*. Retrieved 5 Nov, 2002, from http://construct.haifa.ac.il/~ilangz/femminism_education_and_critical_theory.html

Habermas, J. (1972). *Knowledge and Human Interests, Theory and Practice, Communication and the Evolution of Society* (J. Shapiro, Trans.). London: Heinemann.

Habermas, J. (1984). *The theory of communicative action (Vol 1) reason and the rationalisation of society* (T. McCarthy, Trans. Vol. 1). Boston: Beacon Press.

Habermas, J. (1989). Social Action and Rationality. In S. Seidman (Ed.), *Jurgen Habermas on Society and Politics: A Reader*. Boston: Beacon Press.

Harding, S. (1986). *The Science Question In Feminism*. New York: Cornell University.

Harris, K. (1994). *Teachers: constructing the future*. London: Falmer.

Hedges, W. (1997). *Timeline of Major Critical Theories in US*. Retrieved 5 Nov, 2002, from http://www.sou.edu/english/IDTC/timeline/uslit.htm

Held, D. (1980). *Introduction to Critical Theory*. Berkeley: UCLA Press.

Hirschheim, R., & Klein, H. (1989). Four Paradigms of Information Systems Development. *Communications of the ACM, 32*(10), 1199-1216.

Horsburgh, M. (1996). *Quality, Quality Audit and Where are we going?* (Unpublished discussion paper for academic board). Auckland: Auckland University of Technology.

Howcroft, D., & Truex, D. (2001). Special Issue on Analysis of ERP Systems: The Macro Level. *The DATABASE for Advances in Information Systems, 32*(4).

Howcroft, D., & Truex, D. (2002). Special Issue on Analysis of ERP Systems: The Micro Level. *The DATABASE for Advances in Information Systems, 33*(1).

ISI. (2002). *ISI Web Of Science*. Retrieved 5 Nov, 2002, from http://www.isinet.com/isi/products/citation/wos/

Jennings, L., & Graham, P. (1996). Exposing Discourses Through Action Research. In O. Zuber-Skerrit (Ed.), *New Directions in Action Research* (pp. 49-65). London: Falmer Press.

Kaminski, J. (2002, 6 Aug 2002). *Nursing Informatics, Section 7: WWWSites, Humanism - Critical social theory Sites*. Retrieved 5 Nov, 2002, from http://www.nursing-informatics.com/kwantlen/wwwsites13.html

King, K. (1994). What Counts as Theory? In K. King (Ed.), *Theory in its Feminist Travels* (pp. 1-54). Bloomington and Indianapolis: Indiana University Press.

Klein, H., & Myers, M. (1999). A Set of Principles for Conducting and Evaluating Interpretive Field Studies in Information Systems. *MIS Quarterly, 23*(1), 67 - 93.

Kuhn, S., & Muller, M. (1993). Introduction to the Special Issue on Participatory Design. *Communications of the ACM, 36*(4), 26-28.

Kuhn, T. (1962). *The Structure of Scientific Revolutions*. Chicago: University of Chicago Press.

Mann, S., & Buissink-Smith, N. (2001). What the Students Learn: Learning through Empowerment. *NZ Journal of Applied Computing & IT, 5*(2), 35-41.

McKay, J., & Marshall, P. (1999). *2x6=12, or Does it Equal Action Research?* Paper presented at the Australasian Conference on Information Systems, Wellington.

McKay, J., & Marshall, P. (2001). The dual imperatives of action research. *Information Technology and People, 14*(1), 46-59.

McKernan, J. (1991). *Curriculum Action Research*. London: Kogan Page.

Melrose, M. (1996). Got a Philosophical Match? Does it Matter? In O. Zuber-Skerrit (Ed.), *New Directions in Action Research* (pp. 49-65). London: Falmer Press.

Melrose, M. (2001). Maximising the Rigour of Action Research? Why Would You Want To? How Could You? *Field Methods, 13*(2), 160 -180.

Mill, J., Allen, M., & Morrow, R. (2001). Critical Theory: Critical Methodology to Disciplinary Foundations in Nursing. *Canadian Journal of Nursing Research, 33*(2), 109-127.

Myers, M. (1995). Dialectical hermeneutics: a theoretical framework for the implementation of information systems. *Information Systems Journal, 5*(1), 51-70.

Myers, M. (2000). *Qualitative Research in Information Systems*. Retrieved June 16, 2000, from http://www.auckland.ac.nz/msis/isworld.html

Myers, M., & Young, L. (1997). Hidden Agendas, Power and Managerial Assumptions in Information Systems Development. *Information Technology and People, 10*(3), 224 - 240.

O'Neill, M. (1996). Prostitution, Feminism and Critical Praxis: profession prostitute? . [Online]. Available: http://www.staffs.sc.uk/schools/humanities_and_soc_sciences/sociology/level3/prost3.htm [5 November 2002]. Staffordshire University. *Austrian Journal of Sociology*(Winter).

Orlikowski, W., & Baroudi, J. (1991). Studying Information Technology in Organizations: Research Approaches and Assumptions. *Information Systems Research, 2*(1), 1 - 28.

Reeves, T. (1997). Established and Emerging Evaluation Paradigms for Instructional Design. In C. Dills & A. Romiszowski (Eds.), *Instructional Development Paradigms* (Vol. 1, pp. 163-178). Englewood Cliffs, New Jersey: Educational Technology Publications.

Robinson, H. (1994). *The ethnography of empowerment: the transformative power of classrooom interaction*. Bristol: Falmer Press.

Rocco, T. (1998). Deconstructing Privilege: An Examination of Privilege in Adult Education. *Adult Education Quarterly, 48*(3), 171-184.

Smith, L., Mann, S., & Buissink-Smith, N. (2001). Crashing a Bus Full of Empowered Software Engineering Students. *NZ Journal of Applied Computing & IT, 5*(2), 69 -74.

Stepulevage, L., & Plumeridge, S. (1998). Women Taking Positions Within Computer Science. *Gender and Education, 10*(3), 313 - 326.

Surendran, K., & Young, F. (2001). Teaching Software Engineering in a Practical Way. *NZ Journal of Applied Computing & IT, 5*(2), 75 - 79.

Susman, G., & Evered, R. (1978). An Assessment of the Merits of Scientific Action Research. *Administrative Science Quarterly, 23*(December), 583 - 603.

Switala, K. (1999). *Feminist Critical Theory*. Retrieved 5 Nov, 2002, from http://www.cddc.vt.edu/feminism/cri.html

Taket, A., & White, L. (1993). After OR: An Agenda for Postmodernism and Poststructuralism in OR. *Journal of the Operational Research Society, 44*(9), 867-881.

White, L., & Taket, A. (1994). The Death of the Expert. *Journal of the Operational Research Society, 45*(7), 733-748.

Wilson, F. (1997). The truth is out there: the search for emancipatory principles in information systems design. *Information Technology and People, 10*(3), 187-204.

Yeaman, A. (1994). Special Issue: The Ethical Position of Educational Technology in Society. *Educational Technology, 34*(2).

Zuber-Skerrit, O. (Ed.). (1996). *New Directions In Action Research*. London: Falmer Press.

3

Programming Environments for Novices

Mark Guzdial

The task of specializing programming environments for novices begins with the recognition that programming is a hard skill to learn. The lack of student programming skill even after a year of undergraduate studies in computer science was noted and measured in the early 80's (Soloway, Ehrlich, Bonar, & Greenspan, 1982) and again in this decade (McCracken, Almstrum, Diaz, Guzdial, Hagan, Kolikant, Lazer, Thomas, Utting, & Wilusz 2001). We know that students have problems with looping constructs (Soloway, Bonar, & Ehrlich, 1983), conditionals (Green, 1977), and assembling programs out of base components (Spohrer & Soloway, 1985)—and there are probably other factors, and interactions between these factors, too.

What are the critical pieces? What pieces, if we "fixed" them (made them better for novice programmers), would make programming into a more manageable, learnable skill? If we developed a language that changed how conditionals work or loops, or make it easier to integrate components, would programming become easier? That's the issue that developers of educational programming environments are asking.

Each novice programming environment (or family of environments) is attempting to answer the question, "What makes programming hard?" Each answer to that question implies a family of environments that address the concern with a set of solutions. Each environment discussed in this chapter attempts to use several of these answers to make programming easier for novices.

Obviously, there are a great many answers to the question "What makes programming hard?" For each answer, there are a great many potential environments that act upon that answer, and then there are a great many other potential environments that deal with multiple answers to that question. That's not surprising, since it's almost certainly true that there is no one correct answer to the question that applies to all people.

Not all of these potential environments have been built and explored, however. The field of Computer Science Education Research is too new, and there are too few people doing work in this field. We are still in the stage of the field of identifying *potential* answers to key questions—indeed, even figuring out what the key questions are!

Nonetheless, there *are* many novice programming environments that have been built, and not all can be discussed in a short primer. Instead, this chapter will focus on three families that have been particularly influential in the development of modern environments and in the thinking of the CS Ed research community.

- The *Logo* family of programming environments, that began as an off-shoot of the AI-programming language *Lisp* and spawned a rich variety of novice programming environments.
- The *rule-based* family of programming environments, that drew from both Logo and *Smalltalk-72*, but even more directly, *Prolog*.
- The *traditional programming language* family of novice programming environments, which tried *not* to change the language, but instead provide new student-centered supports for existing programming languages.

The audience for these environments ranges from young school children for the Logo environments to undergraduate university students for some of the traditional programming language environments. In this chapter, the issue of student differences (e.g., age, background, motivation) is simply glossed over. Such a huge simplification is acceptable in this situation because the problem is so hard. No matter what the age of the students, programming is hard to learn. Whether students attempt to learn to program at a young age or at the age of young adults, the tasks and difficulties remain similar. The environments in the sections below are attempting to deal with those challenges at whatever the age of the student audience.

Research Methods in Programming Environments for Novices

How do we know if a programming environment for novices *works*? What do we mean by "works"? Strange as it may seem, there are few good answers to the first question, in part because of the wide variety of answers to the second question.

Consider how one would go about studying the question of what it means for a programming environment to work. Let's say that we define an environment as working well if users can write more lines of program code, in less time, and with fewer bugs than a comparative programming environment. How would we study that question? We might set up two groups to compare, each learning and using a different programming environment. Perhaps the real value of a programming environment does not become clear until one has used it *for a long time*. So, let's plan on each group spending six months to a year working in the given programming environment. Perhaps the real value of a programming environment does not become clear until one has worked *on large problems* with it. So, let's plan on each group working on some large problems (10,000 lines of code? 100,000? 1,000,000?).

You might see why such comparative studies are rare. It is expensive to develop expertise in a programming environment. It is a huge expense to have two teams working on the exact same large problem, just to develop a result about the quality of one environment over the other.

These problems exist at the novice level, too, but are confounded by the fact that learning to program is hard and different programming environments may have strengths at different points in the process. Is the given programming environment particularly strong at helping students get started (e.g., may feature a graphical programming environment that is easy and engaging to use)? Or is it better when the students start struggling with design issues (e.g., may feature tools and visualizations that aid in design)? Are these comparable?

In studying novice programming environments, the question of what it means to "work" may be subsumed by the question of why students are being taught to program at all. If the goal is to teach higher-order thinking skills, then perhaps the research question should focus more on the *effects of* using the programming environment, rather than the *effects with* use of the programming environment (Salomon, Perkins, & Globerson, 1991)

The problem of comparing novice programming environments is related to a problem that the entire field of computer science education has in attempting to study computer science learning. We do not have a strong theoretical base with powerful measurement instruments. We do not really know how one learns to program. What understanding comes first? What are necessary preconditions for successful learning? We know a lot about the problems that students have in learning to program, but we know less about the causes of those problems and what a successful developmental process of computer science learning looks like.

Thus, much of the research in novice programming environments is *ad hoc*. There are not generally accepted research questions nor methods. Perhaps the greatest contributions to be made in this field are not in building yet more novice programming environments but in figuring out how to study the ones we have.

Logo and its Descendants: The Goal of Computational Literacy

Logo was developed in the mid-1960's by Wally Feuzeig and Danny Bobrow at BBN Labs, in consultation with Seymour Papert at nearby MIT. Logo was designed to be "Lisp without parentheses." Lisp was a popular programming language for artificial intelligence programming. Lisp was known for its flexibility and the ease with which data could become program, or vice-versa, making it very easy for programs to manipulate their own components. Lisp was especially good for creating and manipulating representations of knowledge. (See Figure 1 for the family tree of this section.)

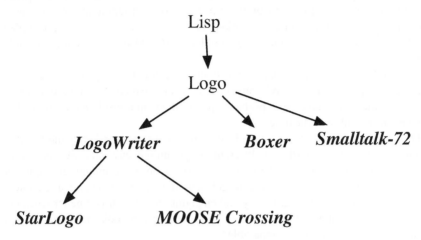

Figure 1. The Logo family of novice programming environments (italics indicate real environments tuned to novices)

The answer to the question of "What makes programming hard?" for the Logo developers was another question. When Logo was first being developed, people didn't *know* that programming was going to be so hard for so many. Programming was still a curiosity, an activity practiced only by the few who had access to the still-rare machines. The Logo developers asked instead "Why should students program?"

The answer to that question was related to Piagetian thinking about learning. Seymour Papert worked for a time in Jean Piaget's laboratory. The goal for Logo was for students to think about their own thinking, by expressing themselves in their programs and then debugging the programs until they worked (Papert, 1980). By debugging their programs, the argument went, the students were debugging their own thinking. Logo proponents argued that Logo then led students to thinking about their own thinking and gain higher-order thinking skills (such as skills at debugging and planning). Papert saw Logo as a tool for immersing students in a "math-land" where they would learn math, as easily as a child learns a foreign language when immersed in a culture where that language is the norm.

Students' early activities with Logo involved mathematical and word games. Logo was especially strong at playing games with language, e.g., creating a Pig Latin generator or a Haiku generator. Later, a robot "turtle" was added to the Logo environment, so that students could control the robot with commands like **forward 1** to move the robot forward a little bit, **right 1** to turn right one degree, and **pendown** so that the robot's pen (literally, a pen attached to the robot) would draw on the surface below the robot as it moved. With the turtle, Logo could be used for graphics, and in reverse, graphics could be used to understand mathematics. When Logo was moved onto graphical user interfaces (the first uses of Logo were on paper-scrolling teletype terminals), the turtle went with it, but as a graphical object that left graphical pen trails on the screen. With turtle graphics (using the turtle to draw), mathematics, and language support in Logo, student programs in Logo *could* range over a broad set of knowledge areas.

However, Logo soon became intimately linked with turtle graphics. For many, Logo was *only* turtle graphics, and any program that offered turtle graphics was consequently some form of Logo. Logo proponents and researchers pointed out that turtle graphics was significant in its complexity and scope. A book by Abelson and diSessa pointed out how much of mathematics could be addressed through the geometry available through turtle graphics, often in interesting and even novel ways (Abelson & diSessa, 1986). Nonetheless, the close relationship between Logo and the turtle led some to believe that Logo was limited to just explorations of geometry.

Logo use grew through the early 1980's. Many different forms of Logo were developed, and many are still available today [1]. Because of its linkage with turtle graphics, Logo was fairly popular in science and math classes, but not much farther in the curriculum. Papert's vision of integration across the curriculum never really happened.

Research on Logo

Early studies of Logo tended to be observational. Those studies focused on how students used Logo in their classrooms. The question of whether Logo really led to increases in higher-order thinking skills were the ones that researchers were most interested in.

Critiques of Logo began soon after Papert's *Mindstorms* was released and the excitement about it grew (Hoyles & Noss, 1992). The study that most often gets cited for demonstrating that Logo did not work was out of Bank Street College showing little or no cognitive benefits of learning to program (Pea & Kurland, 1986). That study, in turn, has received significant criticism (Hoyles & Noss, 1992)—for example, the study had relatively few students involved, and not all variables were adequately controlled. But as the authors themselves pointed out in a later paper (Kurland, Clement, Mawby, and Pea, 1986), they didn't see themselves claiming that Logo was ineffective for gaining higher-order thinking skills. Rather, they noted that *very few of the students learned to program!* The real result of the Bank Street studies was that programming is hard to learn.

In general, studies of the relationship between higher-order thinking skills and programming have never shown any significant correlation. In his review of the many studies exploring such a connection, David Palumbo found that there is very little evidence to claim such a relationship, and certainly none from short (e.g., single semester) experiences learning to program. (Palumbo, 1990). Transfer of cognitive skill from one context (programming) to other problem solving contexts (e.g., tracing directions on a map) is very difficult to achieve (Bruer, 1993). However, there has been a successful study *creating* such transfer.

Sharon Carver, for her Ph.D. thesis (Carver, 1986), developed a cognitive model of what it meant to have transferrable debugging skills, where her example contexts were debugging Logo programs and debugging directions against a map. She then explicitly taught those skills in the context of Logo programming, and tested students for their ability to transfer the debugging skills to the map context (Klahr & Carver, 1988). The students did learn the Logo debugging skills and did demonstrate transfer of those skills. It's important to note, though, that Carver was explicitly *teaching* for transfer—her curriculum was designed around teaching the higher-order thinking skill, not teaching Logo *per se*.

Later studies of learning Logo sought more traditional cognitive skills gain, rather than high-order thinking skills. Idit Harel's dissertation work focused on 4th grade (9–10 years old) students using Logo to develop educational software to teach 3rd grade students about fractions (Harel, 1991). Harel found that the context of teaching others about a subject was motivating for the students and engaged them in thinking hard about the subject (fractions) and their medium (Logo). Her studies comparing her 4th graders to comparison groups of the same age studying the same topics showed significantly greater learning about both fractions and Logo than the comparison groups (Harel & Papert, 1990).

Examples of Logo

The basic commands of Logo could be combined and iterated using recursion or simple looping constructs, like repeat. Executing: **repeat 4 [fd 100 rt 90]** generates:

The syntax of Logo is simple and sparse, like Lisp. It has few special characters. Most of the syntax derives directly from the provided procedures: If you know the procedure, you can figure out how the statements parse. The argument for this kind of syntax was to make it simple for the students to learn the rules and understand the programs.

There are no end line markers (e.g., semi-colons) in Logo. Instead, each procedure (like **fd**) knows how many inputs it needs. The parsing and evaluation are tied tightly in Logo. Code can be contained in lists which are delimited with square brackets ([ ]). Thus **repeat** is a function that takes two inputs: A number of iterations to execute, and a list of code to execute (evaluate) that many times.

We can define procedures to "teaching the turtle" to do something (the language used to explain to children what programming was about). Here is the procedure used for defining the word square to mean the procedure of drawing a square.

```
to square
```

```
repeat 4 [fd 100 rt 90]
end
```

We can now generate the square with **square**. By parameterizing the square procedure, we can draw squares of different sizes.

```
to square :size
repeat 4 [fd size rt 90]
end
```

We can now generate the square with **square 100**.

The syntax for specifying parameters in Logo is drawn from the general syntax for variables. Unlike most programming languages, Logo remained close to its Lisp roots and distinguished between the *value* of the variable and the *name* of the variable. Logo proponents argued that such distinctions improved students' understanding of what the programs were doing. We can see the use of parameterization by exploring the square procedure the way that children were expected to play with squares.

If we move the turtle slightly and then repeat the square procedure, we can get interesting designs. **repeat 100 [square 100 fd 10 rt 30]** generates a figure like this:

Language play is just as natural in Logo. In the below example, the procedure start defines three lists. Each time that **makeOne** is executed, a random (**pick**) element from the list is selected and composed into a sentence (**se**) which is then shown (**show**). The results are sentences like "Mommy runs quickly" and "Daddy jumps high". We see here the syntax for defining a variable (**make**) requires the name of the variable (e.g., **"names**) versus the syntax for accessing the variable (**:names**).

```
to start
make "names [Matthew Katie Jenny Mommy Daddy]
make "verbs [runs eats jumps walks sits]
make "adverbs [high quickly perfectly slowly
peacefully]
end

to makeOne
show (se pick :names pick :verbs pick :adverbs)
end
```

Programming in support of a task

The next step in the evolution of Logo was to consider "What tasks did students want to use programming for?" Or, to build upon the core question of Logo, "What domains did students want to learn about *through* programming?" The first versions of Logo basically offered a programming environment that was a variation of a command line: A graphical area was visible.

The first kind of Logo that really changed the students' programming environment *LogoWriter*. LogoWriter integrated a word-processor, capable of both graphics and text, with a Logo interpreter. From a language perspective, the LogoWriter turtle could now act as a cursor changing letters beneath it, or stamping graphics onto the page. From an environment perspective, LogoWriter felt as much like an applications program as a programming language. Now, students could do language manipulation where they *saw* the language manipulation (as the cursor moved and the words changed), and create programs that constructed mixed text-and-graphics the way that they might in other applications software. LogoWriter took seriously providing task support so that the range of potential domains to explore with programming was as broad as possible. LogoWriter was used by Idit Harel in her thesis studies where she had fourth-graders (10 years old) building software to teach fractions to third-graders (Harel & Papert, 1990).

StarLogo followed along the path of extending the language and the environment to focus on a particular *kind* of task. StarLogo supported exploration of decentralized systems (Resnick, 1997). Mitchel Resnick provided students with not one turtle, but literally thousands of turtles—all running essentially the same, small program. He also introduced the notion of a *patch*, a spot on the screen that could hold state (such as a color) and could run programs, but could not move as a turtle could. By using patches to represent food for ants or wood chips for termites, StarLogo could be used to explore how ants (turtles) gather food or how termites (turtles) create piles, all without coordination but through the power of simple, distributed programs. Resnick's studies were observational—his goal was not to show that StarLogo was better than anything else, but that students *could* learn about decentralized systems using such a tool.

MOOSE Crossing: Practical Logo for Communities

MOOSE Crossing (by Amy Bruckman) again tuned Logo to a particular domain and task, but a social one rather than a scientific or academic task. MOOSE Crossing is a shared, textual, virtual reality. Students sign on to MOOSE with specialized client software and explore a world created by peer students (all under 12 years old)—and extend the world themselves. Students might create specialized rooms where everything said in the room is turned into Pig Latin, or specialized objects like pet dragons that follow their owners around. Students move around, control their world, and interact through Logo-like commands. These commands can then be strung together in procedures such that the dragon "wags its tail" (i.e., displays the words "dragon wags its tail" to all those in the same room of the virtual space) when the dragon is "pet" (i.e., some user in the same room types "pet the dragon.") The turtle is replaced with text describing the student-created world (Figure 2).

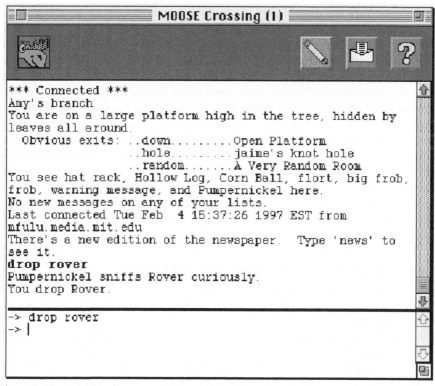

Figure 2. The MOOSE client running on a Macintosh

Bruckman made several changes in the language for MOOSE, based on the experience of years of work in Logo and her concern about making the language accessible to novice programmers. For example, she decided to remove the distinction between the name and the value of the variable. Variables were simply known by name, and whether the value or name was being accessed was determined by context, as in more traditional programming languages. Bruckman used the design principle of "Prefer intuitive simplicity over formal elegance" in make this tradeoff (Bruckman, 1997). She made similar tradeoffs elsewhere in her language. Where **say "Hi there! How is your project coming?"** was acceptable in MOOSE, so was **say Hi there! How is your project coming?** in recognition that students had difficulty with syntax issues like matching quotes.

Bruckman's design principle is an important recognition of what novices understand and what's needed to understand abstractions. The simple observation that students are not the same as experts is common in the cognitive science literature, especially with respect to novice and expert software designers and programmers (Jefferies, Turner, Polson, & Atwood, 1981; Abelson & Soloway, 1985). A programming language or environment feature that may make sense for experts may be confounding for novices. Current learning theory, based on the work by Jean Piaget, suggests that students need concrete experiences before they can understand abstractions on the experiences (Turkle & Papert, 1991). The abstraction of separating names and values may make more sense to experts who have significant experiences than students facing their first programming environment.

Bruckman found that students did learn programming in this environment, supported and motivated by the social context (Bruckman, 2000). Her studies were *ethnographic*—she observed the students, talked with them, and got to know them so that she understood them, but not so that she could compare them with other students using other kinds of systems. Students in MOOSE Crossing were able to communicate with one another, share their creations, and even teach each other to program with no external (e.g., adult) support–a surprising event that she documented in a detailed case study. Her later studies used more quantitative approaches of measuring activity in MOOSE Crossing, and she was able to show that MOOSE was successful at overcoming some gender stereotypes. She showed that girls were just as successful at programming as boys, if the context is motivating (Bruckman, Jensen, and DeBonte, 2001). This was a strong finding in favor of Papert's initial premise that it was what one did with programming that really mattered most.

Smalltalk and Boxer: Extending Beyond Logo

Smalltalk-72, by Alan Kay, Dan Ingalls, Adele Goldberg, and other members of the Xerox PARC Learning Research Group, extended the model of Logo in several ways. Smalltalk was developed along the path to creating the *Dynabook*, a computer whose purpose is to support learning through creation and exploration of the rich range of media that a computer enables (Kay & Goldberg, 1977). Kay agreed with Papert that computers should be used by students for knowledge expression and learning through debugging of those expressions. However, he felt that the computational power provided by Logo was too weak, so he invented *object-oriented programming* as a way of enabling much more complex artifacts to be created in exploration of more complex domains. The command-line metaphors of Logo were too weak for the drawing, painting, and typeset-quality text that Kay felt was critical in order to enable rich media creation, so he and his group literally invented the desktop user interface as we know it today (Figure 3). Within this metaphor, Smalltalk provided a wide variety of programmer tools within the environment, including code browsers, object inspectors, and powerful debugging tools.

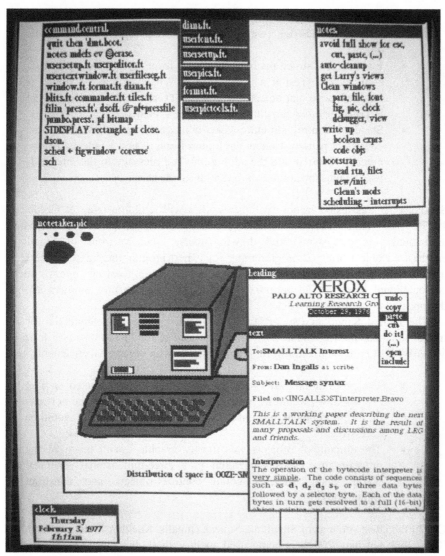

Figure 3. Smalltalk-72, the first system with overlapping windows, icons, menus, and a mouse
pointing device—all designed for the novice programmer.

Smalltalk's syntax is similar to Logo's, in that there are a small number of
grammatical rules and very few reserved syntactic structures. Smalltalk developers
felt that the few number of rules would make it simpler to learn. In Smalltalk, every
statement is of the form <receiverObject> <message>. The focus is on *all
computation arising from message sends*.

Modern Smalltalks have some of that flavor, but are more complicated than the
Smalltalk-72 that was used with children. Nevertheless, the syntax of modern
Smalltalks gives a sense of what it means for all computation to be object-oriented
and message based.

- A statement like **4 printString** means "Send the integer 4 the message **printString** (which returns the text representation of the object)."
- Even control structures have this basic form. (a=b) ifTrue: [Smalltalk beep] means "Test if a is equal to b. Whatever boolean object is returned, send that object the message **ifTrue:**. If the boolean object is **True**, the *block* Smalltalk beep will be executed."
- Standard control structures like while and for follow the same consistent pattern. **1 to: 10 do: [:index | sum := sum + index]** adds the numbers 1 to 10 into the sum. **to:do**: is a message to the integer 1, which takes the arguments 10 and the code block.

Smalltalk-72 was tested in observational studies with children (Kay & Goldberg, 1977). The studies showed children producing amazing programs, including animations, music systems, and drawing tools. Later versions of Smalltalk emphasized object-oriented programming for expert programmers and the desktop user interface for applications software–and de-emphasized ideas about computational literacy. The notion of Smalltalk for novice programmers all but disappeared for perhaps 15 years.

The United Kingdom Open University adopted a form of Smalltalk for its introductory computing course. The Smalltalk used in the OU course is a form of VisualWorks (a direct ancestor of Smalltalk-72), but with several enhancements.

- The programming environment was carefully constructed so that few windows would be on the screen at once and only the portions of the environment relevant to the students' current project would be accessible.
- The domain of projects was mostly graphical with much of the programming projects centered in a three-dimensional graphical world where students could control objects and construct simulations.

One of the latest versions of Smalltalk, Squeak (Ingalls, Kaehler, Maloney, Wallace, Kay, 1997; Guzdial, 2001), is being used again with students, especially younger children. *Squeak* is Smalltalk from the late 1970's minimally updated to run on modern machines, but then augmented with a wide range of new features, especially in support for multimedia. An alternative interface for using Squeak has been implemented *e-toys* that allows for a drag-and-drop tiling-based programming environment. Students literally drag variables, values, and methods from place-to-place to define procedures, mostly to control graphical objects (Figure 4)–and mostly with more complex syntax than in traditional Smalltalk. Like in Bruckman's MOOSE Crossing, the Squeak e-toys interface favors concreteness and ease of use to powerful abstractions. The e-toys interface has been used with success with 10–12 year old students (Guzdial & Rose, 2001).

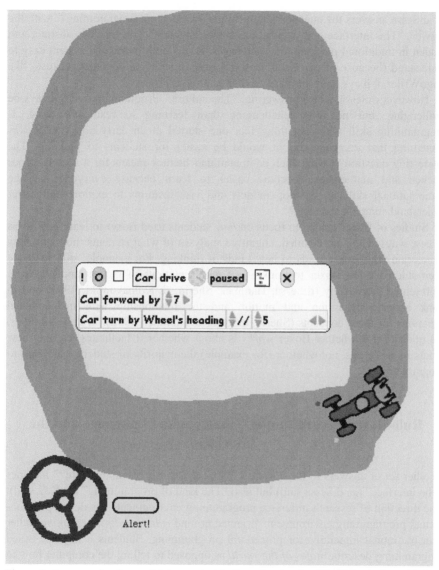

Figure 4. An example of a car-driving project in Squeak using the tiling-based

Andrea diSessa also extended Logo, but in a different direction. Rather than tune it to a specific task, he tried to think about what computation would look like if it were a real *literacy*—as ubiquitous as text reading and writing is today (diSessa, 2001). Boxer was based on a principle of *naive realism*: Every object in the system has an on-screen graphical representation that can be inspected, modified, and extended. For example, variables are not just names in Boxer. Creating a variable creates a named *box* on the screen which corresponds to that variable. Setting the variable's value changes (visibly) the contents of the corresponding box. Changing the contents of the box (with direct manipulation and typing) changes the value of the variable. Lists exist in Boxer, but as list boxes.

diSessa answers the question about "What's hard about programming?" with the answer, "The interface and its relation to the language" Too much is abstract and hidden in traditional programming languages. Boxer both makes the system easy to understand (because of naive realism) and easy to apply to domains because, like LogoWriter, it plays upon similarity to applications software.

However diSessa is also answering, "The culture." Programming will always be challenging, but no more challenging than learning to read and write. If programming skill was something that one started at an early age, and it was something that *everyone* did, it would be easier for students to pick up. The interesting question is what such computational literacy means for a society. Does science and mathematics become easier to learn because everyone has the computational skills to develop models and visualizations to explore and better understand complex concepts?

Studies of Boxer tended to focus on *how* students used Boxer to learn new ideas in new ways. They are detailed, cognitive analyses of what students think and how interaction with Boxer leads to new kinds of thinking. For example, students have been studied using Boxer to think about representations of motion and how it is represented graphically (diSessa, Hammer, Sherin, & Kolpakowski, 1991) and to think about acceleration and motion and the relationship with mathematical equations for these concepts (Sherin, diSessa, & Hammer, 1992). In these contexts, the question for whether Boxer *works* is about whether it facilitates learning new things in new ways, not whether (for example) the majority of students can learn to program in it.

Rule-Based Programming: Changing the Language and the Interface

Another set of answers to the question "What makes programming hard?" includes "The interface" (as diSessa said) but also "The kind of programming" (as Kay said). One direction of research in novice programming environments has developed non-textual programming environments oriented around *rule-based programming* rather than traditional imperative or procedural programming. Students using rule-based programming describe *states of the world* as opposed to telling the computer how to operate upon the world.

Prolog was a popular rule-based programming language, even with novices, even when it simply had a command-line kind of interface. In Prolog, one states facts about the world, e.g., "The factorial of *n* is 1 if *n*=1, and otherwise, it's the *n* multiplied by the factorial of *n*-1." That isn't explicitly telling the computer *how* to get a factorial: It states a definition of factorial, which happens to be complete enough to be executable. That's how Prolog works. Prolog avoids some of the complexities of loops and conditionals with which research shows students have difficulty.

For example, Prolog can be taught facts by simply entering them, e.g.,

```
| parent(tom, bob).
| parent(bob, jim).
```

```
| parent(tom, liz).
```

The database can then be queried, using specific statements or parameterized *patterns*, as in:

```
| ?- parent(tom, bob).
yes
| ?- parent(tom, john).
no
| ?- parent(tom, X).
X = bob ?
```

Note that **X** in the above example is not a variable in the sense of most programming languages. **X** is not a name associated with some data. Instead, X is an *unbound* variable whose binding is found by search in the Prolog logic database.

We can state logical relations on top of these facts, such as the grandchildren of Tom being those Y's who are children of X where Tom is X's parent.

```
| ?- parent(tom, X), parent(X, Y).
X = bob
```

Prolog thus emphasizes the *logic* of programming, without confusing the matter with graphics or even procedures. Programmers in Prolog specify how things are *related* without specifying *what* is to be done–leaving those details to Prolog. For Prolog proponents, Prolog boils programming down to the basic activity of stating logical relations.

Extending Prolog into graphical domains

Some versions of Prolog could be used to generate graphics or other media, but the core of Prolog's descendants are entirely graphical. They have boiled the language down to a *matching rules* representation. The Prolog While some versions of Prolog could be used for graphical and other tasks not strictly textual, other languages carried Prolog into a purely graphical language. These latter languages emphasize applications in a concrete, graphical domain appropriate to creating graphical simulations or videogames. It's a domain motivating for many students, and it lends itself toward providing immediate feedback to students' explorations. Just as we saw in the progression of Logo environments, and in the movement from Smalltalk to the Squeak e-toys interface, the progression of environments after Prolog is toward concreteness and graphical domains.

Stagecast Creator[2] and AgentSheets[3] both explicitly support rule-based programming *and* a different kind of interface for programming. The developers of Stagecast (formerly *KidSim* and *Cocoa*) explicitly aimed to use all that was learned about direct-manipulation interfaces to make the task of programming easier (Smith, Cyber, & Spohrer, 1994). In both of these tools, the user defines rules that describe how the state of the world should change if particular conditions are met.

For example, consider the AgentSheets simulation of a train (in Figure 5). The rule appears on the right of Figure 5 (from Repenning, Ioannidou, & Zola, 2000).

The rule states that if the train is on the track on the left of the track, the train should move forward onto the right of the track.

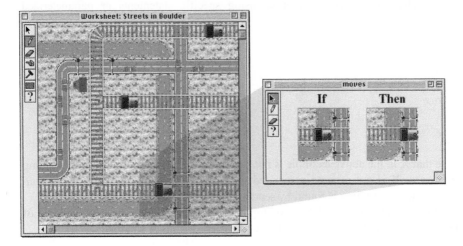

Figure 5. An AgentSheets simulation of a train

Both AgentSheets and Stagecast Creator are most often used for building simulations or video games (Figure 6). The graphical nature of rules lend themselves to the kinds of motion and manipulations that many videogames provide. Both AgentSheets and Stagecast Creator support non-graphical rules, as well. For example, a more complex *if* condition can lead to a set of *then* actions, including sounds and setting variable values.

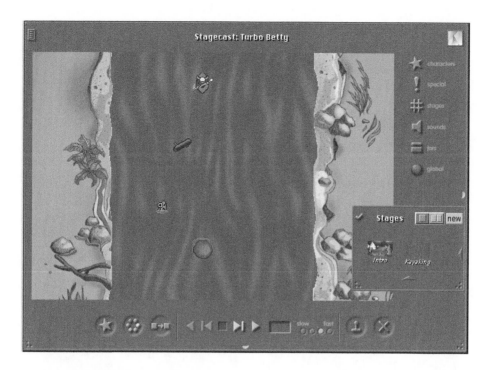

Figure 6. Stagecast Simulation

Both are used extensively in educational settings. Stagecast Creator has been used by kids to build a variety of videogames, including several for Internet competitions. AgentSheets has been used in some quite remarkable simulations for exploring social studies, e.g., simulations of how peaceful protests can become riots, in a StarLogo-like fashion (Repenning, Ioannidou, & Philips, 1999). In general, the focus in these kinds of environments continues to be on what can be learned using these kinds of tools (Repenning, 1994).

ToonTalk[4] by Ken Kahn is explicitly influenced by the work of Seymour Papert, but it follows the rule-based and non-text model of Stagecast Creator and AgentSheets. ToonTalk takes the model of programming-as-videogame much further than these other two environments. In ToonTalk, a student's program explicitly manipulates *characters* who, in turn, manipulate data and structures of data which appear as Lego bricks (Figure 7). The rendering of ToonTalk is exceptionally high-quality: The look-and-feel is as nice as a high-end videogame (Figure 8).

Figure 7. ToonTalk characters manipulate Lego-like data

Figure 8. ToonTalk running has multiple characters and assembled data elements, all rendered in beautiful quality

ToonTalk gives the same answers to "What makes programming hard?" as Stagecast Creator and AgentSheets, but it provides some additional ones.

- ToonTalk is concerned with making it obvious *who* is doing what a program commands. Agency is made visible through its characters.
- Like Boxer, ToonTalk includes naive realism in that everything is visible. What's more, ToonTalk provides the metaphor of Lego to make clear how virtual objects are, literally, *assembled*.
- ToonTalk takes great pains to make sure that its execution has the same realism as high-end videogames. For example, ToonTalk (like StarLogo) provides a high degree of concurrency—things happen at

once, just as they do in the real world. Kahn believes that this makes it easier for students to understand and develop in ToonTalk.

Toontalk has been used in classroom contexts for observational studies as long as three years (Kahn, 1999). The goal has been to understand *what* students learn with ToonTalk, and in particular, what hard concepts are now approachable because of ToonTalk.

Putting a New Face on an Old Language: Programming for Future Programmers

Still another answer to the question of "What makes programming hard?" is to say, "It's not the programming language at all." There's an argument that using idiosyncratic or *ad hoc* programming languages decreases student motivation, since the programming skills developed can't be used elsewhere. Perhaps the answer is "It's the programming environment—it needs to support learning the skills of expert programmers." This answer is probably most relevant to those students studying computer science as a potential profession, since they are clearly motivated to learn existing programming languages. Some researchers have argued that it's also relevant to students studying programming to learn problem-solving skills (Soloway, 1986). Environments that act upon this answer emphasize teaching design skills and *scaffolding* (providing additional support that students need but experts don't (Blumenfled, Soloway, Marx, Krajcik, Guzdial, & Palincsar, 1991)) students to use traditional languages. In particular, the more complex syntax of traditional programming languages was viewed as a stumbling block for novice programmers, so much of the scaffolding was aimed at relieving syntax complexity.

Most of this work was done when Pascal was the dominant programming language in schools (Figure 9). These environments had in common support for *structured editing* and design support. Structured editing refers to how the text of the program is manipulated. Rather than simply typing the textual program, structured editors support specification of elements (e.g., from menus) and the completion of *placeholders* that fill in the details of the program.

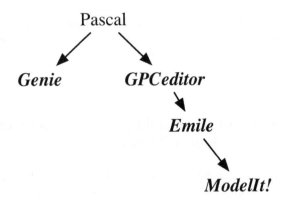

Figure 9. Family of novice programming environments based on supporting traditional
    programming languages

Probably the largest effort to create structured Pascal editors for students was the
Genie effort at Carnegie-Mellon University (Miller, Pane, Meter, & Vorthmann,
1994). Over some ten years, several different Genie editors were created. All of
them provided structured editing support. For example, a user might choose a for
loop to be inserted into her code. The loop would be inserted with placeholders
identifying where additional pieces needed to be specified (Figure 10), which could
be completed by selecting placeholders and making menu selections. Genie also
provided visualizations in its debuggers so that diSessa's principle of naive realism
was used to facilitate debugging (Figure 11).

```
For $control-variable$ :- $start-value$ $direction$ $end-value$ Do
    Begin
        $statement$
    End
```

Figure 10. A Genie for loop with placeholders to-be-completed

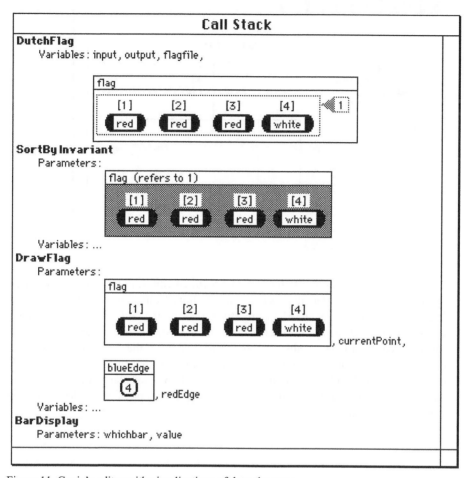

Figure 11. Genie's editor with visualizations of data elements

The Genie developers also realized that part of students' problem with programming was in figuring out how to start and how to move forward to completion—again, lessons learned through observation, rather than any comparative studies or even ethnographic analysis. The Genie developers decided that the issue was that students lacked design skills, as others were deciding around the same time (Soloway, 1986). Genie provided design views of programs that explicitly encouraged students to see their programs as sets of components that they were assembling (Figure 12).

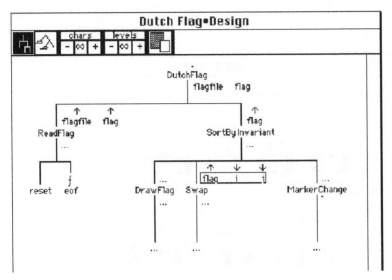

Figure 12. Genie's design view

The GPCeditor was another Pascal-based structured editor, like Genie, but it started from the design view (Guzdial, Kinneman, Walton, Hohmann, & Soloway, 1998). Rather then choose language elements, students using the GPCeditor specified their *goals*, and then selected *plans* from a *Plan Library*, which were instantiated as *code* (thus, Goal-Plan-Code editor). Figure 13 shows the students goal-plan decomposition in the upper left, their Plan Library in upper right, the actual code (with the selected code corresponding to the highlight goal-plan) in lower left, and the hierarchy of goal-plans in the lower right (overlapped with the execution window of a running program).

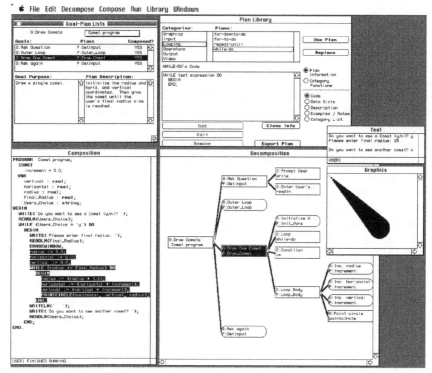

Figure 13. GPCeditor: Goal-Plan-Code editor

The GPCeditor was notable for its evaluation effort. It was used in secondary schools for several years, and findings suggest that students did develop design skills that transferred from GPCeditor to more traditional Pascal programming environments (Guzdial et al., 1998). The methodology used was an artifact analysis—looking at what the students did in GPCeditor, and then what the same students did in traditional Pascal, and identifying similarities. However, it's not clear that GPCeditor made the task any *easier*.

Successors to GPCeditor followed the same progression as Logo, in that they provided support for specific kinds of tasks that might themselves be motivating to students. *Emile* supported students building hypermedia (based on HyperCard) programs that simulated physical systems (Guzdial, 1995), using a goal-plan-code structure to ease the editing of the (fairly) traditional programming language. The study of Emile was mostly an observational study but it was paired with pre-test and post-test clinical interviews (a carefully scripted interview to address particular points). That study showed that students did learn a lot about physics through construction of their simulations, but less about programming.

A still later successor to GPCEditor, ModelIt! (Soloway, Jackson, Klein, Quintana, Reed, Spitulnik, Startford, Studer, Eng, & Scala, 1996), supported simulation of systems, but moved away from programming to a direct-manipulation system that supported specification of relationships between simulation factors as manipulation of graphical and textual statements–a similar progression as from Prolog to its successors.

Summary: Trends and Future Directions

In the over forty years of programming environments for novices that we skimmed over here, it is clear that the research community has only started to address the question of "What's hard about programming?" We can identify several trends in the research.

First, there is a clear trend toward a more traditional language syntax. The powerful abstractions of the earliest programming languages for novices have been de-emphasized in favor of programming languages that are easier to read, perhaps with support to make them easier to write. The syntax of Bruckman's MOOSE Crossing and of Squeak's e-toys are more complex than the Logo and Smalltalk ancestors–the parser has to work harder to figure out what's going on and the number of rules for interpreting what's going on grows. Why should a more complex language be easier for students to understand? It's not obvious, but we can make some conjectures.

- Novice programmers may not actually think about the process of interpretation or compilation. To think about *how* the language is understood by the computer is a level of abstraction that is beyond the novice.
- Natural language *is* complex and ambiguous. As the programming languages reflect natural complexity, they may become easier for the novice to understand, even though the mathematical and logical clarity decreases.
- Finally, it's not clear how much computer science influences novice programmers, especially when the students are in secondary or post-secondary education. In one of our recent introductory courses at Georgia Tech, 83% of the student respondents to a survey claimed that they had no previous programming experience. But when interviewed, most professed to using Logo or Basic in elementary school, or looking through books or magazines about programming at some point. Computer science is part of modern culture, and some of the predominant programming cultures have some influence on students.

Second, there is another trend toward relevance and applicability. Students want to work on computational artifacts that have meaning for them, e.g., that are interesting and relevant. Research studying how students relate to computer science and why they don't stay with it has shown that a *lack* of meaning and relevance are key issues for students' distaste for computer science (e.g., Margolis & Fisher, 2002). The turtle graphics of Logo and logic databases of Prolog have faded in favor of building video games, text and graphical virtual realities, and simulations.

Third, related to the trend toward relevance and applicability is a trend toward environments and tasks that give students immediate feedback on their work. Students want to "pet Rover" and see their car turn and move their robot. The issue is more than just being motivating to pet the dog and see the graphical reaction. If the students' programs work, they see immediate and understandable responses: Rover wags his tail, the car turns, and the robot moves. Contrast this with a program to sort a list of names or compute the n-th digit of π and then printing the result. The student has to analyze the result and decide if it really is sorted or really does reflect the correct digit of π. When students are dealing with the complexity of learning programming, they don't want or need to deal with subtle shades of correctness–they want it to be right or wrong, so that they can correct it and move on.

There are a great many issues that have scarcely been addressed in novice programming environments, and many more that arise due to new technologies. Moore's Law constantly changes the technological scenery, thus changing what's possible for novice programming and what novices can do with their new programming skills.

- For example, we know that today's children use processors all the time, but not on the desktop—in their handheld videogames, cellphones, and Palm Pilots. Is programming too locked to the desktop? What does programming mean on these smaller devices that might be more motivating (relevant and applicable) for students to manipulate?
- In some new work that we're undertaking at Georgia Tech, we have students learning to program by manipulating media, such as changing pixel color values to generate grayscale or by moving sound samples to modify sounds. Students start our writing relatively short programs that loop over the pixels or samples–but the loops loop *many* times. Four seconds of sound at low-quality is some 88,000 samples. If something goes wrong in the students' programs, printing variable values doesn't make sense. Nobody wants to wade through 88,000 lines of output. Traditional debuggers and breakpoints aren't useful, either–most debuggers are too confusing for novices, and nobody wants to hit the button *Continue from Breakpoint* 88,000 times. How do we help students to understand their programs in this kind of domain?
- The interaction between the programming environment and language has just started to be addressed. Computer speed is such that the parser can be running all the time, while editing, so that immediate feedback is offered during program entry. How do we use this computational power to scaffold the novice programmer?
- And as mentioned at the start of this paper, we still have a weak theoretical basis and few techniques for measuring learning in computer science. That leaves us with few techniques for studying and comparing novice programming environments.

Finally, it may be that "What makes programming hard?" is not the most fruitful question to ask. Perhaps we don't know yet what programming really is or what it could be (as diSessa suggests in his book (2001)). Perhaps we don't know yet what students would really want to learn programming for.

Designing programming environments for novices is a fascinating field that we have only just started to explore. There is a great many more questions and answers to explore, and some wonderful environments yet to build and try in that exploration. The progress in the field is toward making programming more interesting, more relevant, and more powerful for students. The research opportunities could hardly be broader, and promise enormous potential impact.

References

Abelson, H. and diSessa, A. A. (1986). *Turtle Geometry: The Computer as a Medium for Exploring Mathematics*. MIT Press, Cambridge, MA.

Adelson, B. and Soloway, E. (1985). The role of domain experience in software design. *IEEE Transactions on Software Engineering*, SE-11(11):1351–1360.

Blumenfeld, P. C., Soloway, E., Marx, R. W., Krajcik, J. S., Guzdial, M., and Palincsar, A. (1991). Motivating project-based learning: Sustaining the doing, supporting the learning. *Educational Psychologist*, 26(3 & 4):369–398.

Bruckman, A. (2000). Situated support for learning: Storm's weekend with rachael. *Journal of the Learning Sciences*, 9(3):329–372.

Bruckman, A., Jensen, C., and DeBonte, A. (2001). Gender and programming achievement in a cscl environment. In Stahl, G., editor, *Proceedings of the 2002 Computer Supported Collaborative Learning conference*, pages In–Press. University of Colorado at Boulder, Boulder, CO.

Bruckman, A. S. (1997). *MOOSE Crossing: Construction, Community and Learning in a Networked Virtual World for Kids*. PhD thesis, MIT.

Bruer, J. T. (1993). *Schools for Thought: A Science of Learning in the Classroom*. MIT Press, Cambridge, MA.

Carver, S. M. (1986). *Transfer of LOGO debugging skill: Analysis, instruction, and assessment*. PhD thesis, Carnegie-Mellon University, Pittsburgh, PA.

diSessa, A. (2001). *Changing Minds*. MIT Press, Cambridge, MA.

diSessa, A. A., Hammer, D., Sherin, B., and Kolpakowsi, T. (1991). Inventing graphing: Meta-representational expertise in children. *Journal of Mathematical Behavior*, 10:117–160.

Green, T. R. G. (1977). Conditional program statements and comprehensibility to professional programmers. *Journal of Occupational Psychology*, 50:93–109.

Guzdial, M. (1995). Software-realized scaffolding to facilitate programming for science learning. *Interactive Learning Environments*, 4(1):1–44.

Guzdial, M. (2001). *Squeak: Object-oriented design with Multimedia Applications*. Prentice-Hall, Englewood, NJ.

Guzdial, M., Konneman, M., Walton, C., Hohmann, L., and Soloway, E. (1998). Layering scaffolding and cad on an integrated workbench: An effective design approach for project-based learning support. *Interactive Learning Environments*, 6(1/2):143–179.

Guzdial, M. and Rose, K. (2001). Squeak, open personal computing for multimedia. Prentice-Hall, Englewood, NJ.

Harel, I. (1991). *Children Designers: Interdisciplinary Constructions for Learning and Knowing Mathematics in a Computer-Rich School*. Ablex, Norwood, NJ.

Harel, I. and Papert, S. (1990). Software design as a learning environment. *Interactive Learning Environments*, 1(1):1–32.

Hoyles, C. and Noss, R. (1992). *Learning Logo and Mathematics*. MIT Press, Cambridge, MA.

Ingalls, D., Kaehler, T., Maloney, J., Wallace, S., and Kay, A. (1997). Back to the future: The story of squeak, a practical smalltalk written in itself. In *OOPSLA '97 Conference Proceedings*, pages 318–326. ACM, Atlanta, GA.

Jefferies, R., Turner, A. A., Polson, P. G., and Atwood, M. E. (1981). *The processes involved in designing software*. Lawrence Erlbaum Associates, Hillsdale, NJ.

Kahn, K. (1999). Helping children to learn hard things: Computer programming with familiar objects and actions. In Druin, A., editor, *The Design of Children's Technology*. Moran Kaufman, New York.

Kay, A. and Goldberg, A. (1977). Personal dynamic media. *IEEE Computer*, pages 31–41.

Klahr, D. and Carver, S. M. (1988). Cognitive objectives in a logo debugging curriculum: Instruction, learning, and transfer. *Cognitive Psychology*, 20:362–404.

Kurland, D., Clement, C., Mawby, R., and Pea, R. (1986). Mapping the cognitive demands of learning to program. In Pea, R. and Sheingold, K., editors, *Mirrors of Minds*, pages 103–127. Ablex, Norwood, NJ.

Margolis, J. and Fisher, A. (2002). *Unlocking the Clubhouse: Women in Computing*. MIT Press, Cambridge, MA.

McCracken, M., Almstrum, V., Diaz, D., Guzdial, M., Hagan, D., Kolikant, Y. B.-D., Laxer, C., Thomas, L., Utting, I., and Wilusz, T. (2001). A multi-national, multi-institutional study of assessment of programming skills of first-year cs students. *ACM SIGCSE Bulletin*, 33(4):125–140.

Miller, P., Pane, J., Meter, G., and Vorthmann, S. (1994). Evolution of novice programming environments: The structure editors of carnegie-mellon university. *Interactive Learning Environments*, 4(2):140–158.

Palumbo, D. B. (1990). Programming language/problem-solving research: A review of relevant issues. *Review of Educational Research*, 60(1):65–89.

Papert, S. (1980). *Mindstorms: Children, computers, and powerful ideas*. Basic Books, New York, NY.

Pea, R. and Kurland, D. (1986). On the cognitive effects of learning computer programming. In Pea, R. and Sheingold, K., editors, *Mirrors of Minds*, pages 147–177. Ablex Publishing, Norwood, NJ.

Repenning, A. (1994). Programmable subsrates to create interactive learning environments. *Journal of Interactive Learning Environments*, 4:45–74.

Repenning, A., Ioannidou, A., and Phillips, J. (1999). Collaborative use and design of interactive simulations. *Proceedings of Computer Supported Collaborative Learning Conference at Stanford (CSCL'99)*.

Repenning, A., Ioannidou, A., and Zola, J. (2000). Agentsheets: End-user programmable simulations. *Journal of Artificial Societies and Social Simulation*, 3(3).

Resnick, M. (1997). *Turtles, Termites, and Traffic Jams: Explorations in Massively Parallel Microworlds*. MIT Press, Cambridge, MA.

Salomon, G., Perkins, D., and Globerson, T. (1991). Partners in cognition: Extending human intelligence with intelligent technologies. *Educational Researcher*, 20:2–9.

Sherin, B., diSessa, A. A., and Hammer, D. (1992). Dynaturtle revisited: Learning physics through collaborative design of a computer model. *Interactive Learning Environments*, 3(2):91–118.

Smith, D. C., Cypher, A., and Spohrer, J. (1994). Kidsim: Programming agents without a programming language. *Communications of the ACM*, 37(7):55–67.

Soloway, E. (1986). Learning to program = learning to construct mechanisms and explanations. *Communications of the ACM*, 29(9):850–858.

Soloway, E., Bonar, J., and Ehrlich, K. (1983). Cognitive strategies and looping constructs: An empirical study. *Communications of the ACM*, 26(11):853–860.

Soloway, E., Ehrlich, K., Bonar, J., and Greenspan, J. (1982). What do novices know about programming? In Badre, A. and Schneiderman, B., editors, *Directions in Human-Computer Interaction*, pages 87–122. Ablex Publishing, Norwood, NJ.

Soloway, E., Jackson, S., Klein, J., Quintana, C., Reed, J., Spitulnik, J., Stratford, S., Studer, S., Eng, J., and Scala, N. (1996). Learning theory in practice: Case studies of learner-centered design. In Trauber, M., editor, *CHI96 Conference Proceedings*, pages 189–196. Vancouver, British Columbia, Canada.

Spohrer, J. C. and Soloway, E. (1985). Putting it all together is hard for novice programmers. In *Proceedings of the IEEE International Conference on Systems, Man, and Cybernetics*, volume March. Tucson, AZ.

Turkle, S. and Papert, S. (1991). *Epistemological pluralism and the revaluation of the concrete*, pages 161–192. Ablex, Norwood, NJ.

Notes

1http://el.media.mit.edu/logo-foundation/products/software.html
2http://www.Stagecast.com
3http://www.agentsheets.com
4http://www.toontalk.com

4

Research on Learning to Design Software

W. Michael McCracken

Introduction

This chapter overviews the research methods and techniques used to understand how software designing is learned. The chapter emphasizes the methods of cognitive science for understanding design and design learning. The chapter begins with an overview of design learning research and follows with an explanation of software design and how it is different than programming. The next section outlines research methods employed in studying software design.

Research in software design learning is quite sparse. Historically, researchers have studied expert designers or compared novice and expert designers, c.f., (Adelson & Soloway, 1988; R. Guindon, 1987; Jeffries, Turner, Polson, & Atwood, 1981; Sutcliffe & Maiden, 1992). The emphasis of those studies was to generate cognitive models or explanations of design knowledge and skill. They sometimes compared expert and novice design behavior. The problem with those studies is that they often look at novices and experts as bipolar entities, with little description of the development of the skills and knowledge that allows a novice to move toward being an expert.

Other design fields similarly lack research specifically aimed at learning to design. In engineering and architecture much has been written about novices and experts or comparisons of both, c.f.,(Radcliffe & Lee, 1989; Ullman, Dieterich, &

Staufffer, 1988),but little has been written about the learning processes associated with gaining expertise in design.

The recent workshop on Design Knowing and Learning gathered researchers interested in design learning (software, engineering, architecture and industrial design) to present recent work on design learning (C. Eastman, McCracken, & Newstetter, 2001). Though the workshop and resultant book hoped to foster research in design learning, as of yet, there is still very little being published on the topic. There are some exceptions. Schon in the 1980's studied architectural design learning in the studio (Schon, 1987). More recently, Atman has published results associated with engineering design learning, c.f., (Atman & Bursic, 1996, 1998; Atman & Turns, 2001).

The point of the introductory comments is neither to dissuade someone from researching design learning nor to lament the lack of research. In fact, the field is open to researchers who wish to gain a better understanding of design learning and possibly apply those results in the classroom. This chapter reviews the methods, problems and difficulties, as well as the rewards of learning about design learning.

Design and Learning to Design

There are three research areas that contribute directly or indirectly to studying design learning.

- **Design as a Cognitive Process** Design can be described and understood from many perspectives. Theories of cognition inform most current research in design, since designing is a type of problem solving and problem solving has been primarily studied as a cognitive process.
- **Learning as a Cognitive Process** Learning can similarly be described and understood as a cognitive process. Though there are many theories of learning theories this chapter will focus on learning to solve problems.
- **Experts and Novices as Cognitive Differentiators** Many studies of problem solving rely on novice - expert comparisons. Since learners are at some point novices aspiring to be experts, this chapter will rely on expert and novice behavioral descriptions of software design, as well as other design fields.

The next three subsections will cover the areas.

Design as a Cognitive Process

Definitions of design are almost as numerous as the number of designers. Designing is described as any problem solving activity that results in the creation of an artifact or a plan for generating an artifact. Simon describes it as:

Design on the other hand is concerned with how things ought to be, with devising artifacts to attain goals (Simon, 1981).

As such, design is a process composed of the activities a person or persons engage in to solve a design problem. The design process begins with understanding the problem and ends with generating satisficing solutions of the problem. The design activities are dependent on the domain the designer is operating in. Though that definition is quite narrow and at the same time almost vacuous, it will serve as an entree' to a more explicit set of definitions.

Describing design activities in more detail improves the definition. One source of information on the activities of designers is the research results from other design fields. The literature of all design fields is useful to software design researchers, but at times the terminology is confusing. To aid the beginning design researcher, Table 1 compares the terminology of different design fields primary design activities.

Table 1: A Comparison of Design Terminology

	Software Design	Engineering Design	Architectural Design
Problem Formulation	Requirements Spec	Functional Spec	Project Spec or Program
Conceptual Design	Software Architecture	Functional or System Block Diagram	Elevations and Models
Preliminary Design	Class Models and Interaction Diagrams	Functional Decompositions or Sub-systems	Plans, Critical Sections and Critical Material Specs
Detailed Design	Program Specs	Circuit or Component Designs	Construction Specs and Construction Detail Drawings
Implementation	Programming	Manufacturing	Construction
Evaluation	Testing	Testing	Evaluation

The basic activities of software design are: requirements gathering and analysis, architectural design, detail design, implementation and testing. These activities have been observed or prescribed in many forms over the years. Most textbooks on software design include activity lists and describe them in detail, c.f.,(Braude, 2001; Pressman, 2001; Sommerville, 2001). For example, the requirements development activity described by Sommerville as the requirements engineering process includes (Sommerville, 2001):

- Requirements elicitation and analysis
- System models
- Requirements specification including user and system requirements
- Requirements validation
- Requirements documentation

The activity descriptions can serve as a framework for teaching or learning software design but can also be misused, as descriptions can become prescriptions. For example, a student learning requirements analysis may blindly follow the steps of analysis without understanding why they are doing it, that is, the reason for requirements analysis is to understand the problem adequately to be able to construct a solution for it.

In the past, activity descriptions were used to prescriptively teach design. Design methodologists proposed specific steps and representations for software design. The methods prescribed specific actions, such as, understand the problem before solving it, which was translated into, write the requirements specification before designing the solution. Prescriptive methods of teaching design have merit. A novice designer who has no idea of how to attack a problem can begin with prescribed steps. Though it is hard to argue with that suggestion, it can be misinterpreted as, never start designing the solution until the problem is completely understood. For example, in the 1970's the U.S. Department of Defense adopted a set of standards that forced completion of requirements before allowing design to proceed, completion of design before implementation could proceed, etc. Those prescriptive methods closely mimicked the design methods movement in other design fields (N. Cross, 1984), though trailing it in time by almost a decade. In a classroom context, the prescriptions can be interpreted by students as a task to be performed to achieve a grade (Newstetter, 1998). The methodological approach emphasized products of activities, such as a requirements specification, rather than the cognitive processes employed by the designer to generate the information that became the product. There are many reasons for the adoption of prescriptive design methods, but they were primarily due to the lack of understanding of how complex design problem solving is which came from strictly observing designing. Thus, the cognitive problem solving actions of a designer were shrouded in mystery.

Researchers began studying designers as cognitive agents rather than document producing machines as early as the 1960's (C.M Eastman, 1969). The earliest substantive studies of software design occurred in the late 1970's and early1980's, c.f., (Jeffries et al., 1981). These studies of software design behavior described it as iterative and opportunistic (R. Guindon, 1987) and a cognitively intensive process dependent on domain knowledge, computation knowledge and skill, and control mechanisms to manage the design process (Soloway & Ehrlich, 1984). These descriptions were often presented as cognitive models of design and represented as schemas or other information processing based models (Adelson & Soloway, 1988; R. Guindon, 1987; Jeffries et al., 1981). All of these studies relied on one characteristic of designing, the enactment of problem solving skill. Guindon, et al's study is an example of this work (R Guindon, Krasner, & Curtis, 1987). Their model of software design was based on an opportunistic control schema versus balanced (strict breadth first) decomposition. They gave their subjects problems that were more complex than used in prior studies, to understand how more realistic problems were actually solved by experts. Their hypothesis was that balanced or strict breadth first decomposition was not used by experts when they were confronted with problems they had not previously solved. Their research method was protocol analysis of their subjects. They observed decomposition behavior in their subjects until they became stuck in some particular area, at which time the subjects moved to non-decomposition methods to attempt and evaluate partial solutions. They developed a model of designing that accommodated opportunistic behavior. The model relied on a set of inputs consisting of the problem environment, or "all the objects, properties, behaviors, and constraints in the world that are relevant to the design of a solution" (R. Guindon, 1987), pp. 19. It also included a knowledge element, which consisted of representations and processes in the problem domain and solution domain (R. Guindon, 1987) pp. 19.

Goel and Pirolli gave a more explicit definition of design. They proposed that all designing could be described as a set of invariant activities (invariant across design problem spaces) (Goel & Pirolli, 1992). A designer's behavior is therefore affected by the set of invariants. Since a design researcher often observes designers designing, the invariants give clues to the expected types of behavior (at least from expert designers). Goel and Pirolli's invariants are quite important to the issues of studying design learning as well as differentiating design problem solving from problem solving. Cognition as information processing is the basis of most studies of design. In other words, models of design are goal-seeking processes (Newell & Simon, 1972), and Goel and Pirolli depend on that basis for their study. Goel and Pirolli's invariants are summarized in the following list.

- **Distribution of Information** The start state and goal state of design problems are initially poorly specified. Thus, design problems are ill-structured (Simon, 1973). Ill-structured problems can be characterized as having loosely defined start states and goal states, have a large number of constraints that need to be resolved as a part of solving the problem, and numerous acceptable or satisficing solutions (Voss & Post, 1988)
- **Nature of Constraints** Design problems have nomological and social, political, etc., constraints. These constraints effect the designer's solution options as well as complicating the problem. For example, a design is incorrect if it is economically infeasible or socially unacceptable, even if it technically solves the problem.
- **Size and Complexity of Problems** Design problems are typically large and complex. Real design problems take days, months or years to solve, not an hour or two. Multiple domain experts are frequently needed to solve design problems.
- **Component Parts** There is little information in design problems to guide the designer in their decomposition. Design problems are not adequately described to dictate lines of decomposition and are primarily decomposed based on the designer's experience. For example, a problem of designing a stock market prediction system has little inherent decomposition paths within the problem.
- **Interconnectivity of Parts** Design problem components are not logically interconnected. The components of design problems (sub-problems) don't have logical interconnections. Most design problems have some defined external connections, but few defined internal connections until the problem solution is developed.
- **Right and Wrong Answers** Design problems only have better or worse answers, not right and wrong answers. This is an important aspect of studying the artifacts of designing. How does an evaluator determine the quality of the design?
- **Input/Output** Inputs to design problems are the desires of the person(s) who desire the artifact. Outputs are the specifications of the artifact.

- **Feedback Loop** During design there is no real feedback from the world on the design. Any feedback on the design process is generated by simulated or generated by the designer. For example, the users of a system have difficulty assessing a system until it is fielded and in operation.
- **Cost of Errors** Errors propagated through a design can be costly to correct. Frequently solutions have errors when they are fielded and don't manifest themselves until then. Those errors can be catastrophic or trivial, but the cost of correction is not a function of their impact. For example, a missile failed to launch because an error caused a computer to reboot during launch.
- **Independent Functioning of the Artifact** The artifact must function independent of the designer. A design solution must stand-alone from the designer. For example, a program should not require its designer to operate it for the user.
- **Distinction between Specification and Delivery** The specification of an artifact and its delivery are distinct. Software design blurs the distinction between specification and construction. Since material choices are of minimal consequence in designing software, the construction aspects focus on language and platform choice.
- **Temporal Separation between Specification and Delivery** An artifact is specified prior to its actual delivery. Even with the blurring of specification and construction in software design, a system is generally specified and reviewed prior to coding. This results in a time lag between the specification and delivery of the system.

Goel and Pirolli's invariants of the design task environment can be summarized as: design problems are typically ill-structured, don't have a well defined start state, goal state, and transition process, are big and complex, cannot rely on prescriptive approaches to solving them, must fit into the world and function without the designer being present, and are likely expensive to repair once fielded. Software design obviously exhibits the invariant characteristics of designing as described by Goel and Pirolli.

Software designers employ cognitive processes that are similar to the processes used by designers in other fields. Software design is differentiated from other design fields by stating that the designer typically constructs the delivered product and those products are not physical objects. In addition, software designers require domain knowledge as well as computational knowledge (Detienne, 2002). The computational knowledge required to design software is more than programming knowledge.

Is Software Design Different than Programming?
The prior subsection made no comments about programming other than the obscure reference to specification versus delivery. Software design differs from many design fields because the designer constructs the artifact to be delivered versus subcontracting construction to skilled laborers. A machinist or construction worker does not require the education of a programmer. Yet, there is a difference between programming and software design. The differences will be stated parochially and you can decide whether these definitions match or collide with yours. [1]

Programming is the process of converting an abstract specification of a problem solution into a concrete artifact, a program. If we include the problem solving process that led to the specification, we can still differentiate programming from design by recognizing that a program is *domain less* or *decontextualized.* That is, the specification of a program has lost its identity as a domain artifact. The program is a model of a computational artifact. For example, a designer is developing a web site that catalogs all of the results of art auctions. One of the requirements of the system is that users can search for items using various parameters, such as, artist, date, medium, subject, etc. The designer has completed a detailed design for the search requirement and has generated a data model that offers efficient searching and has defined the algorithms for searching. That detailed design is essentially domainless. It is easy to retrieve the necessary domain knowledge by looking back at the architecture or the requirements, but the detailed design itself has little context of the original problem. The designer's concerns are focused on searching and data structures, regardless of the problem (noting that the requirements for the web site are what caused the search problem to exist). This problem and its solution could occur in many different contexts.

Software design is embedded in a domain. Software design problems are generally not in the domain of computation, but in finance, medicine, entertainment or transportation. Studying software design therefore requires the researcher to understand, differentiate, and sometimes combine, the designer's domain skill and knowledge as well as the designer's computational skill and knowledge.

The difference between programming and software design, as previously stated, is somewhat arbitrary. If a student is solving a particular problem, is it programming or design? Listed below are two differentiators to clarify how a researcher can view designing as compared to programming.

- Programming is the delivery of the product of software design. Software design can be studied with or without the subject writing a program.
- Programming is the solution of relatively well structured problems that have a single or small set of defined answers. Design is the solution of ill-structured problems that have only better or worse answers.

In summary, the schemas of design rely on the schemas of programming to solve problems. The schemas of programming can generally be mapped to different problem spaces once their relationships are understood and made.

Learning as a Cognitive Process

This section will briefly cover the terminology and views of cognitive science that are relied upon to study design learning. A complete overview or explanation of cognitive science and the learning sciences are contained in many books and articles that are too extensive to even list as a bibliography. A good starting point for general learning theory is (Berliner & Calfee, 1996). A recent book by Detienne is a good reference for those interested in pursuing research in software design (Detienne, 2002).

The basis of cognitive learning theory is to understand why or how a person learned or why they did not learn. Since designing can be looked at as a type of problem solving, a researcher should also be interested in why or how a person solved a problem in a particular way. Prior research in problem solving and learning to solve problems is available to the design learning researcher, but as described in Section the previous section, design problems have some different characteristics than *normal* problems. Those differences make the study of design learning more difficult than the study of problem solving.

Cognitive scientists often rely on a basic model of cognition as an information processor (Newell & Simon, 1972). The information processing model of cognition has been applied extensively in problem solving beginning with the early work of Newell and Simon to the current work of Anderson and others (Anderson & Lebiere, 1998; Newell & Simon, 1972). The ability to model human problem solving has supported the modeling of design cognition. That is, models are constructed that may explain or partially predict the behavior of designers or possibly design learners.

Researchers applying information processing models of design frequently rely on a representation based on schemata. A schema is an abstraction (similar to an object in OO design) that can be instantiated by assigning values to its variables. Schemas can be realized as computational models or can be used as a non-computational model of memory. [2] A problem solver could have a schema for adding numbers, and the instance of the schema would be the assignment of the values to be added and the resultant answer. Designers have complex schemas that include sub-schemas, just as a balancing checkbook schema would include an adding numbers sub-schema. A designer of web applications could have a schema for screen layout that includes sub-schemas for controlling buttons or pull downs. The button sub-schema could have sub-schemas that include the programs or program outlines for event handlers for the buttons. Expert designers have many schemas learned from experience, and extensive pattern matching and analogy generating processes that support problem solving. Experts also have control schemas that are used to manage the overall design process. Schemas can also contain experience or episodic knowledge. As such, a designer could have a schema of a pre-solved problem that only needs to be retrieved versus enacted to solve the problem. One of the questions asked by design learning researchers is how are the schemas constructed, modified and tuned as a person learns designing (Norman, 1982)? [3]

Many of the questions asked in learning research are applicable to design learning research. Questions such as, why couldn't the student solve a particular problem, or why did the student use an inappropriate solution, or why did the student omit a particular part of the problem from their solution, when placed in context, are questions of how people learn to design software.

Even though there have been few studies of software design learning, a study by Guindon, et al, to define automation aids to support designing can be related to learning design. Their study revealed breakdowns of designers when solving the N-Lift problem (R Guindon et al., 1987). The study described the behavior of three professional software designers and noted the problems (breakdowns) they had solving the problem. Their categorization of breakdowns illuminated knowledge and skill deficiencies of practicing designers that were due to varying learning deficiencies, that is, a lack of knowledge and/or cognitive limitations of their subjects. The breakdowns seen by Guindon, et al, are similar to the behavior of novices.

The breakdowns and the category of knowledge or cognitive limitations are listed below:

Breakdowns due to Lack of Knowledge

1 A lack of specialized design schemas. Their subjects did not have pre-existing schemas for all of the sub-problems of the design.

2 A lack of a meta-schema about the design process. Some of their subjects did not have an overall control process for the design activities. The meta-schema is about design not about the problem.

3 Poor prioritization of issues leading to poor selection of alternative solutions. Some of their subjects were unable to manage their resources (primarily time) as a function of the priority of issues that needed resolution.

Breakdowns due to Cognitive Limitations

1 Difficulty in considering all the stated or inferred constraints in refining the solution. Some of their subjects failed to evaluate their proposed solutions against the constraints of the problem, and as a result their solutions failed to satisfy those constraints.

2 Difficulty in performing mental simulations. Their subjects were unable to rely on simulation of partial or complete solutions to aid in evaluation of the correctness of their design.

Breakdowns due to Cognitive Limitations and Lack of Knowledge

1 Difficulty in keeping track of sub problems whose solution had been postponed. Their subjects were at times unable to remember to return to deferred sub-problems.

2 Difficulty in expanding or merging solutions from sub-problems into a complete solution. The expansion and integration of sub-problem solutions requires knowledge of computer science and a history of the design process, which some of their subjects did not have. (R Guindon et al., 1987)

Novices and experts have been referred to in the previous sections. The following subsection will describe in more detail what differentiates a novice from an expert.

Experts and Novices as Cognitive Differentiators

The reason for studying experts is they can be used as *benchmarks* of behavior. The benchmarks can be used to illuminate behavioral differences when compared to novices. For example, a researcher may observe an expert spending more time understanding a problem than a novice. The observed behavior of the expert could indicate that she was developing strategies and plans for solving the problem prior to embarking on a solution. The comparative behavior of the novice could indicate he was moving directly into solving the problem without any strategy.

In the late 1960's and early 1970's there were studies of what differentiates experts from novices (Chase & Simon, 1973; deGroot, 1966). These studies and many others looked at problem solving in physics, math and writing. Some of the studies contrasted expert and novice behavior c.f.,(M.T.H Chi, Glaser, & Farr, 1988; M.T.H Chi, Glaser, & Rees, 1982; Ericsson & Smith, 1991). As previously stated, many of the studies of software designing described novice, expert or novice/expert behavior.

Two of the expert/novice studies of software design found results that illuminated some of these issues. Jeffries, et al, compared novice, pre-novice and expert software designers (Jeffries et al., 1981). They found that novices when compared to experts did not consider or evaluate alternatives, were inconsistent in their approaches, and dedicated little time to problem formulation. Adelson and Soloway conducted four experiments with subjects and problems as the variables (Adelson & Soloway, 1985). Their novice/expert findings were similar to Jeffries, et al.

These and other studies found that expert software designers had an ability to see and represent problems at a more principled level than novices. That means an expert can directly relate problems (domain specific) and solutions (computational specific) as one. For example, an expert logistics software designer would have the ability to see an inventory addition problem and its potential solution as a single schema.

The following concepts from expertise research are important for studying design learning.

- Expertise is domain dependent. Experts at designing logistics software will likely have problems designing a digital library. This is due to a lack of domain knowledge and skill, not due to a lack of computational skill and knowledge.
- Experts become experts through practice and experience. An expert has a broad range of experiences in designing software. Those experiences support the experts' retrieval processes as well as their ability to construct analogies of similar situations.
- Expert's knowledge can be represented as complex schemas that are composed of sub-schemas. Expert's sub-schemas can be adapted to solve different problems.
- Programming skill and knowledge are parts of the adaptable sub-schemas. For example, a logistics software expert may recognize the similarity of sub-problems in a digital library problem and map the solutions onto the new problem space.

The results from the Jeffries, et al study maps directly to the general expertise concepts (Jeffries et al., 1981). Their novice subjects failed to retrieve relevant domain knowledge. Their novice subjects had little experience in how to effectively decompose the problem, had minimal ability to solve sub-problems, and failed to rely on relevant information from the classroom.

The description of the cognitive activities of design, the relevant aspects of learning to solve problems, and the difference between experts and novices forms a basis for studying design learning. The following section briefly overviews some of the techniques used to study learning to design.

How To Conduct Research on Design Learning?

The answer to the question asked in the section head is that it depends. In general, researchers in design learning are interested in understanding the behavior, that is, the cognitive processes, employed in designing. Behavioral research of this type is difficult to perform by reviewing the results of designing. For example, a subject or subjects produce a sequence diagram for a software system and the researcher describes the subject's behavior by analyzing the diagram. At best, even with large samples, the researcher can only generate inferences of what the subjects were thinking when they generated the sequence diagram.

The evolution of design research has moved from case studies and observation of designing to protocol studies and in some cases controlled laboratory studies. Cross describes the evolution of design research as having progressed through four stages:

prescription of an ideal design process; description of the intrinsic nature of design problems; observation of the reality of design activity; and reflection on the fundamental concepts of design (N. Cross, 1984).

This progression is reflected in the methods employed in design research. The research methods moved from observation and thought experiments to more formal protocol analysis. As researchers gained a better understanding of design cognition, they were able to see that there were many common domain-independent behaviors of designers, see for example, (N Cross, 2001; Goel & Pirolli, 1992; Zimring & Craig, 2001).

As with all research, the questions being asked define the research methods to be employed. The types of questions a software design learning researcher might ask are:

- Why do students not see the relation between design patterns and their designs?
- Why do students start coding before they understand the problem?
- Why do students trained in procedural design methods have difficulty learning object oriented methods?

A researcher must observe designing as it occurs to answer those questions.

The approach to studying design cognition is to conduct studies where the subjects are designing and record their activities and if possible determine the subject's internal cognitive processes used in designing. The results of the studies can be used to generate a cognitive model of design c.f.,(Adelson & Soloway, 1985), develop an understanding of the misconceptions the subjects have c.f.,(Newstetter & McCracken, 2001), relate external (written down) and internal (mental) representations c.f.,(C.M. Eastman, 2001), or other components of the cognitive processes of designing.

This section will focus on three techniques available to researchers who wish to study design learning. A more thorough analysis of research methods for studying design behavior is in (Craig, 2001). The three techniques are:

In Situ Observation

The researcher observes and records the actions of the subject or subjects. The researcher is able to observe the externalized activities of the subject or subjects. Researchers can observe sketching, referencing, note taking, and manipulation of external objects. That information is used to construct understandings or descriptions of the behavior of the designer. This is the easiest method to use, but it does not uncover the internalized cognitive processes that were not manifested as external operations, other than through inference. In situ observation allows the researcher to see what, but not why. As a point of clarification, in situ observation is not synonymous with ethnographic methods (Handwerker, 2002).

Observation is primarily a technique to discover problems. For example, a researcher may observe students designing and see they are frequently referring to programming manuals. That observation may prompt him to ask them why they are referencing programming manuals when they are supposed to be solving a problem. Their response could be that they hoped the programming manual had some sample solutions to their problem. The researcher may infer from that observation and dialogue that students rely on pre-existing examples to support designing and plan some type of study to confirm that hypothesis, similar to the study by Chi, et al with physics students (M.T.H. Chi, Bassok, Lewis, Reimann, & Glaser, 1989).

In conclusion, much of the criticism of early research in design was due to researchers constructing models of design based on observation without understanding the cognitive processes underlying the observed behavior.

Retrospective Interviews

The researcher observes the subject designing. At the completion of the design session, the researcher interviews the subject for recollections of what they were thinking in particular situations during the design session. Another means of retrospective analysis is to video tape the design session and play it back for the designer as the questions are asked. The playback is intended to prompt the subject's recall of thoughts. Retrospective interviews require the researcher to be able to observe external actions of the subject and generate relevant questions that expose internal thoughts that influenced the behavior. It also requires the subject to remember what they were thinking at the time in question.

Retrospective interviews are less exploratory than observation. A researcher has developed a hypothesis of behavior under certain conditions and is attempting to observe that behavior. For example, the researcher in the observation example has decided to construct a study to see how examples influence a student's ability to design. The researcher has constructed a model or hypothesis of how students will recall and use previously learned examples as the basis of the study. During the design task, the researcher notes that the students appeared to be trying to force previous examples into the solution of the design problem without success. During the retrospective interview, the researcher asks the student to recall what she was thinking when she was trying to use a linear search function on an un-ordered list. That is one of many questions that may be asked retrospectively by the researcher.

There are numerous issues that have to be reconciled or accepted in using retrospective techniques. The subject may not remember what they were thinking at the time in question. This could lead to incorrect responses or no responses. The subject may omit important details of their thoughts, since at the end of the experiment, they know how to solve the problem, and as a result, may overlook many of their thoughts when they were struggling to solve the problem. The researcher must also not allow her perceptions to cloud the process. If the researcher believes that the subject should solve a problem in a particular manner, and allows that to guide their observations, the results may be erroneous.

Protocol Analysis

The researcher trains the subject to think aloud, in other words, talk about what they are thinking as they are designing. The subject is exposing their internal thoughts, as they occur, to the researcher so that they can be recorded and analyzed.

Protocol analysis is one of the primary methods of studying design cognition, though not without its critics, c.f., (N Cross, Christiaans, & Dorst, 1996). Protocol analysis has been used since the 1930's, but more recently has been described in detail in (Ericsson & Simon, 1993; Someren, Barnard, & Sandberg, 1994). Protocol analysis was primarily used to study problem solving and was naturally applied to studying designers. A recent study by Cross lists 36 articles on design behavior that used protocol analysis (N Cross, 2001).

The actual taking of a verbal protocol is relatively easy. As in the previous two techniques, a subject is given a problem to solve. The choice of problem is obviously dependent on the nature of the study. For example, a subject may be given a problem that is similar to previously solved problems or completely different than previously solved problems. The difficulty with protocol analysis is in the analysis.

Typically, a researcher transcribes the design session and divides the subject's verbal comments into segments. The segment size is dependent on the type of analysis to be performed. The type of analysis to be performed is dependent on the goals of the study. For example, Ullman studied mechanical engineering processes with the objective of generating a more detailed model of designing than previously done (Ullman et al., 1988). The granularity of the speech segments in Ullman's study was at the level of each utterance.

The coding of the verbal comments in a protocol study is dependent on the granularity of the study. A psychological model of the expected behavior of the subject determines the granularity. The psychological model is typically a procedural model generated from the tasks to be performed in solving a problem and a psychological theory of problem solving (Someren et al., 1994).

A psychological model explicitly relates properties of the task and hypotheses about human problem solving to (verbal) behavior (Someren et al., 1994).

The procedural model can be explicit or abstract as a function of the research question, for example, if the researcher is attempting to understand errors generated in design versus studying the impact of existing knowledge on design behavior.

Protocol analysis depends on the psychological model to organize the data collected in a protocol. Other than issues of granularity, the raw data from a protocol can be analyzed using different psychological models or with different objectives by recoding the data. Noting that the problem must support the particular analysis.

The actual coding of the protocol is onerous. For example, a researcher is trying to compare the design skills of engineers after their first year with engineers after their fourth year. A psychological model of design behavior has been generated that lists the expected procedural steps taken by the subjects. That list would include, problem formulation, generation of sub-problems, analysis of documentation, sketching, etc. The researcher must evaluate each cognitive step of the subject and determine what category the subject's behavior fits into, or in some cases does not fit into. A first year student may begin their design process by specifying materials. The researcher's job is to recognize that behavior, though erroneous, and code it correctly.

The coding process is also subject to error. Inter-rater reliability tests should be performed to ensure the accuracy of the coding process.

Though difficult and time consuming, protocol analysis, also known as, verbal protocol analysis, and think-aloud protocol analysis, is a powerful research technique for exposing the cognitive processes of designing. Table 2 compares the methods.

Anyone familiar with qualitative methods knows there are many other techniques omitted from this list, some of which are used to study design. The three techniques described above are the most commonly used in design cognition research with protocol analysis dominating.

Table 2: A Comparison of the Research Methods

	In Situ Observation	Retrospective Interviews	Protocol Analysis
Difficulty	Easy - Observe and record subject behavior	Harder – Requires researcher to analyze activities and generate questions that may expose subjects thoughts during the activity	Harder - Requires researcher to transcribe subjects utterances and code them against a model.
Sample Size- Sample sizes are typical. Some studies have exceeded these sizes	1 to 20	1 to 10	1 to 5
Findings	A description of behavior. Possible inferences about cognitive processes.	Descriptions of cognitive processes employed at particular instances. May not be able to completely describe behavior due to gaps in information.	Models or descriptions of behavior can be developed.
Issues of applying method	Observation is typically a method of discovering problems rather than answering them. Observation is subject to the Hawthorne effect.	Subjects may not remember or remember correctly what they were thinking during the design. Worse, they may construct remembrances based on what they think the researcher wants.	Subjects thought processes can be interfered with by verbalizations. Subjects may be non-verbal. Subjects need to be trained.
Generalizability of Results	Unlikely	Unlikely	Unlikely

Studying design learning is not easy and is time consuming. Design learning researchers are trying to understand the externalized and internalized behavior of designers as they solve problems. The research methods outlined in this section could be classified as *laboratory* experiments. In other words, the subjects would not necessarily be designing in their natural environment. As in formal experiments, qualitative researchers are trying to reduce the number of confounds or uncontrolled variables in their studies. For example, protocol analysis focuses on the behavior of an individual designer solving a problem constructed by the researcher. If designers were studied in their natural environment (that is, actually practicing design), techniques such as protocol analysis would be very difficult, if not impossible to use.

Studying Designers in their Natural Environment

The cognitive stance of this chapter is based on theories of cognition as structures of information and processes that support the understanding of concepts and the skills of problem solving, reasoning, etc. Most studies of design, designers and design learning have used the research methods of cognitive science that were developed to uncover and understand the individual human mind, in particular, the problem solving ability of the human mind. Rather than viewing cognition as a disembodied information processor, cognitive science researchers are now recognizing that:

> *knowledge is distributed in the world among individuals, the tools, artifacts, and books that they use, and the communities and practices in which they participate* (Greeno, Collins, & Resnick, 1996).

This is referred to as the situative/pragmatist-sociohistoric perspective of cognition or simply situated cognition.

There is a growing body of knowledge from researchers studying situated cognition, see for example, (Hutchins, 1995; Latour & Woolgar, 1986; Lave, 1988; Lave & Wenger, 1991; Resnick, Levine, & Teasley, 1991). To study situated cognition requires the researcher to study cognition *in the wild* (Hutchins, 1995), that is, study the subjects as they are participating in a community of practice. Community of practice is a term used to describe the culture, the practices and the norms of behavior of a group of people that are engaged in some type of collective cognitive activity. The rationale for studying situated cognition is based on the idea that all cognition is influenced by the culture and the environmental components of the culture that surrounds a community of practice, and to understand it requires that it be studied in its natural form. For example, an architect's cognitive processes are influenced by the people the architect interacts with (customers, fellow designers, engineers or other design specialists), the artifacts and materials the designer is interacting with, and the culture of architecture, the culture of the architectural firm and the culture of the customer. To study architectural design in a situative context would require the researcher to be able to study it in its natural state, not under *laboratory* conditions.

The researchers studying situated cognition have primarily come from anthropological and sociological fields. Their research methods support the study of cultures and rely on ethnography or similar techniques. To date, there has been little situated research of software design, but there have been some studies of large projects that offer some insights into the behavior of software designers collaboratively designing (Curtis, Krasner, & Iscoe, 1988; Curtis & Walz, 1990). Those studies were of the case study form, but still offered interesting results for design researchers on the dynamics of collaborative software design.

The intent of this section was to give a brief overview of the next level of research in design learning. If we really want to understand how software designing is learned we must also study designers in their communities of practice. The traditions of cognitive science have not failed us; rather they have given us the opportunity to understand cognition well enough to realize that we must also understand how external factors influence the cognition of the individual. There is much more to be learned from the cognitive view of designing as well as the situated view.

Summary

This chapter has outlined designing and how to study design learning. With the exception of the previous section, the chapter focused on design as a cognitive activity. The emphasis was on software design, though studies in other design fields can be relied on for experimental ideas as well as relevant results. The over-emphasis of the differences between programming and software design was used to illuminate the issues of solving ill-structured versus structured problems. A grounding in the solution of structured problems (most programming problems) generates the lower level schemas used by designers, but the higher level schemas and meta-cognitive skills of designing are necessary to successfully solve typical design problems.

Designing software is not flashes of brilliance that cannot be explained, nor is it something that is taught as a set of steps to be followed in a prescribed manner. Designing software is a creative process, but the creativity relies on a designer's ability to see similarities and explore the relations of those similarities.

The job of design learning researchers is to understand how the cognitive processes of designers are formed or malformed and generate interventions that support learning to design.

References

Adelson, B., & Soloway, E. (1985). The Role of Domain Experience in Software Design. *IEEE Transactions on Software Engineering, 11*, 1351-1360.

Adelson, B., & Soloway, E. (1988). A Model of Software Design. In M. T. H. Chi, R. Glaser & M. J. Farr (Eds.), *The Nature of Expertise* (pp. 185-208). Hillsdale, NJ: Lawrence A. Erlbaum.

Anderson, J. R., & Lebiere, C. (1998). *The atomic components of thought*. Mahwah, NJ: Lawrence Erlbaum Associates.

Atman, C. J., & Bursic, K. M. (1996). Teaching engineering design: Can reading a textbook make a difference? *Research in Engineering Design, 7*(7), 240-250.

Atman, C. J., & Bursic, K. M. (1998). Documenting a process: The use of verbal protocol analysis to study engineering student design. *Journal of Engineering Education.*

Atman, C. J., & Turns, J. (2001). Studying engineering design learning: Four verbal protocol studies. In C. Eastman, M. McCracken & W. Newstetter (Eds.), *Design Knowing and Learning: Cognition in Dsign Education* (pp. 37-62). Amsterdam: Elsevier.

Berliner, D., & Calfee, R. (1996). *Handbook of Educational Psychology*. New York: Macmillian.

Braude, E. J. (2001). *Software Engineering: An Object-Oriented Perspective*. New York: Wiley.

Chase, W. C., & Simon, H. (1973). Perception in Chess. *Cognitive Psychology, 4*, 55-81.

Chi, M. T. H., Bassok, M., Lewis, M., Reimann, P., & Glaser, R. (1989). Self-explanations: How Students Study and Use Examples in Learning to Solve Problems. *Congnitive Science, 13*, 145-182.

Chi, M. T. H., Glaser, R., & Farr, M. J. (Eds.). (1988). *The Nature of Expertise*. Hillsdale, NJ: Lawreance Erlbaum Associates.

Chi, M. T. H., Glaser, R., & Rees, E. (1982). Expertise in problem solving. *Advances in the Psychology of Human Intelligence, 1*.

Craig, D. (2001). Stalking Homo Faber: A Comparison of Research Strategies for Studying Design Behavior. In C. Eastman, M. McCracken & W. Newstetter (Eds.), *Design knowing and learning: Cognition in design education*. Amsterdam: Elsevier.

Cross, N. (1984). *Developments in design methodolgy*. New York: Wiley.

Cross, N. (2001). Protocol analysis for studying design and other emperical methods. In C. Eastman, M. McCracken & W. Newstetter (Eds.), *Design Knowing and Learning: Cognition in Design Education*. England: Elsevier.

Cross, N., Christiaans, H., & Dorst, K. (Eds.). (1996). *Analysing Design Activity*. Chichester, England: Wiley.

Curtis, B., Krasner, H., & Iscoe, N. (1988). A field study of the software design process for large systems. *Communications of the ACM, 31*, 1268-1287.

Curtis, B., & Walz, D. (1990). The psychology of programming in the large: Team and organizational behavior. In J.-M. Hoc, T. R. G. Green, R. Samurcay & D. Gilmore (Eds.), *Psychology of Programming*. London: Academic Press.

deGroot, A. (1966). Perception and memory versus thought: Soome old ideas and recent findings. In B. Kleinmuntz (Ed.), *Problem Solving*. New York: Wiley.

Detienne, F. (2002). *Software Design - Cognitive Aspects*. London: Springer.

Eastman, C., McCracken, M., & Newstetter, W. (2001). *Design Knowing and Learning: Cognition in Design Education*: Elsevier.

Eastman, C. M. (1969). On the analysis of intuitive deign processes. In G. Moore (Ed.), *Emerging techniques in environment design and planning*. Cambridge, MA: MIT Press.

Eastman, C. M. (2001). New Directions in Design Cognition: Studies of Representation and Recall. In C. Eastman, M. McCracken & W. Newstetter (Eds.), *Design knowing and learning: Cognition in design education*. Amsterdam: Elsevier.

Ericsson, K. A., & Simon, H. A. (1993). *Protocol Analysis: Verbal Reports as Data*. Cambridge, MA: MIT Press.

Ericsson, K. A., & Smith, J. (1991). *Toward a general theory of expertise: Prospects and Limits*: Cambridge University Press.

Goel, V., & Pirolli, P. (1992). The Structure of Design Problem Spaces. *Cognitive Science, 16*, 395-429.

Greeno, J., Collins, A., & Resnick, M. (1996). Cognition and Learning. In D. Berliner & R. Calfee (Eds.), *Handbook of Educational Psychology* (pp. 15-46). New York: Macmillian.

Guindon, R. (1987). *A Model of Cognitive Processes in Software Design: An Analysis of Breakdown in Early Design Activities in Individuals* (No. STP-283-87): MCC.

Guindon, R., Krasner, H., & Curtis, B. (1987). Breakdowns and processes during the early activities of software design by professionals. In G. M. Olson, S. Sheppard & E. Soloway (Eds.), *Empirical Studies of Programmers: Second Workshop* (pp. 65-82). Norwood, NJ: Ablex.

Handwerker, W. P. (2002). *Quick Ethnography*. New York: Altimira Press.

Hutchins, E. (1995). *Cognition in the Wild*. Cambridge, MA: MIT Press.

Jeffries, R., Turner, A. A., Polson, P. G., & Atwood, M. E. (1981). The Processes Involved in Designing Software. In J. Anderson (Ed.), *Cognitive Skills and Their Acquistion*. Hillsdale, NJ: Lawrence Erlbaum Associates.

Latour, B., & Woolgar, S. (1986). *Laboratory Life: The Construction of Scientific Facts*. Princeton: Princeton University Press.

Lave, J. (1988). *Cognition, in practice*. Cambridge: Cambridge University Press.

Lave, J., & Wenger, E. (1991). *Situated Learning: Legitimate Peripheral Participation*. Cambridge: Cambridge University Press.

Newell, A., & Simon, H. (1972). *Human Problem Solving*. Englewood Cliffs, N.J: Prentice-Hall.

Newstetter, W. & McCracken, M (2001). Novice Conceptions of Design: Implications for the
    Design of Learning Environments, In C. Eastman, M. McCracken & W. Newstetter
    (Eds.), *Design knowing and learning: Cognition in design education.* Amsterdam:
    Elsevier.Newstetter, W. (1998). Of Green Monkeys and Failed Affordances: A Case
    Study of a Mechanical   Engineering Design Course. *Research in Engineering Design,
    1998*(10), 118-128.
Norman, D. A. (1982). *Learning and Memory.* San Francisco: W.H. Freeman.
Pressman, R. S. (2001). *Software Engineering: A Practitioner's Approach.* Boston: McGraw
    Hill.
Radcliffe, D., & Lee, T. (1989). Design methods used by undergraduate engineering students.
    Design Studies, 10(4), 199-207.
Resnick, L., Levine, J., & Teasley, S. (1991). Washington, DC: American Psychological
    Association.
Schon, D. A. (1987). *Educating the Reflective Practitioner.* San Francisco: Josey-Bass.
Simon, H. A. (1973). The Structure of Ill-Structured Problems. *Artifical Intelligence, 4*, 181-
    201.
Simon, H. A. (1981). *The Sciences of the Artificial.* Cambridge, MA: MIT Press.
Soloway, E., & Ehrlich, K. (1984). Empirical studies of programming knowledge. *IEEE
    Transactions on Software Engineering, 10*(5), 595-609.
Someren, M. W. V., Barnard, Y. F., & Sandberg, J. A. C. (1994). *The Think Aloud Method: A
    Practical Guide to Modelling Cognitive Processes.* London: Academic Press.
Sommerville, I. (2001). *Software Engineering.* Harlow, UK: Addison-Wesley.
Sutcliffe, A., & Maiden, N. (1992). Analysing the novice analyst: Cognitive models in
    software engineering. *International Journal of Man-Machine Studies, 36*, 719-740.
Ullman, D. G., Dietterich, T. G., & Staufffer, L. A. (1988). A Model of the mechanical design
    process based on empirical data. *Artificial Intelligence in Engineering and
    Manufacturing, 2*(1), 33-52.
Voss, J. F., & Post, T. A. (1988). On the Solving of Ill-Structured Problems. In M. T. H. Chi,
    R. Glaser & M. J. Farr (Eds.), *The Nature of Expertise.* Hillsdale, NJ 1988: Lawrence
    Erlbaum Associates.
Zimring, C., & Craig, D. (2001). Defining design between domains: An argument for design
    research ala carte. In C. Eastman, M. McCracken & W. Newstetter (Eds.), *Design
    knowing and learning: Cognition in design education.* Amsterdam: Elsevier.

Notes

[1]Definitions and interpretations are often the culprit when disagreement occurs. For
example, Detienne uses the term programming as the overarching term to describe
all aspects of software development (Detienne, 2002).. Other people use the terms
more strictly, as design is the process of transforming a requirement specification
into a solution specification.

[2]There are numerous debates in the cognitive science community on whose model is
the best and which is the most accurate representation of human behavior.
Nonetheless, schemas will be used as a representation only to avoid the debate, since
most design researchers recognize the difficulty of building computational
enactments of design behavior.

[3]A researcher may choose not to rely on information processing models of cognition,
or schemas as a representation of those models. Regardless, one of the fundamental
questions of design learning is how does a person learn to design.

5

Learning to Program: Schema Creation, Application, and Evaluation

Robert S. Rist

Introduction

Expertise in programming is developed over many years of writing code to solve a series of harder and harder problems. A programmer does not gradually get smarter as time goes on; it is the knowledge that changes. An expert has seen, built, and remembered many solutions, but the solution code cannot be what is stored in the expert's memory. First, it is too large. A complete solution is built up from smaller parts and it is these parts that are combined in new ways to build a new program. Second, it is too concrete. An expert programmer can easily write the "same" program in two different languages, so the code itself cannot define the common structure that is stored in memory.

A plan schema is an abstract solution to a common programming problem; it stores a plan to achieve a common goal. A schema is abstract, so it can be applied in many situations. Schemas fit together, so small schemas are used to build large schemas. A novice builds a basic plan by linking a series of actions together and coding them as statements in a programming language. That plan can then be abstracted and stored in memory as a unit of programming knowledge, a plan schema. The schema can be retrieved from memory and used to build a new, larger solution. This learning mechanism does not change with expertise; the expert just uses larger and more abstract pieces of knowledge. Novices have few, small, concrete, isolated, and fragile schemas, where experts have many, large, abstract, connected, and robust schemas.

There are many ways to solve even a simple problem, however; some of these are good solutions, and some are not so good. Whenever there is a choice in design, a design rule is used to make that choice. Novices do not evaluate their solutions because they know few choices and have few design rules. Experts constantly evaluate their solutions as they are built. A second type of programming knowledge is a set of design rules that leads to the creation of good solutions and the storage of good schemas.

There are many ways to design a solution. A novice starts to write code immediately after reading the problem statement, with no prior planning. A professional programmer plans out a solution at many levels of abstraction before writing a single line of code. An expert spends time on meta-planning, on working out the best way to design a solution for a new problem.

This paper describes how people learn to program in three parts. The first part (sections 2 to 6) presents a model of learning by doing that creates schema knowledge, and shows how this model plays out in the domain of programming. The second part (sections 7 and 8) shows how there are many choices in program construction, and presents a set of design rules that support different choices and lead to different solution structures and schemas. The third part (sections 9 and 10) describes how changes in the form and structure of knowledge lead to the different types of behavior seen at different levels of expertise.

Learning by Doing

Expertise in problem solving is created by solving problems, lots of problems of different types and sizes over a long time. Books can provide principles, facts, and examples, but these isolated pieces must be connected during the process of problem solving. This is the basis of learning by doing (Anzai & Simon, 1979): knowledge is connected during the process of problem solving, stored in memory, and later retrieved to solve a similar problem. Learning to program is thus best understood by analyzing the process of program construction.

A design protocol records the order in which pieces of knowledge are combined to build a solution. A single protocol identifies one set of units and links used by one designer for one problem, and a set of protocols shows the common chunks of knowledge and the links between them. Protocol studies of novice problem solving show a consistent pattern that may be described as backward search from the goal. A problem provides a goal to achieve and a set of facts or inputs that are known to be true. In the domain of physics, for example, the problem might be to find the final speed of an object sliding down a ramp, given the weight of the object and the angle and length of the ramp (Larkin, 1981). A novice starts with the goal (the final speed) and finds a formula in memory or in a book that calculates the speed. The inputs to that formula now become goals to achieve, and new formulas are retrieved and linked. The process repeats until a set of formulas are found that can be solved from the facts given in the problem.

A unit of knowledge in the physics domain is a formula; each formula can be seen as a single action with a set of inputs and a single output. The linked actions define a plan to achieve the goal, and problem solving finds and connects these actions backward from the goal. When the plan is complete, the values in the problem are bound to the input variables and the plan is executed forward from the given facts to the final goal. As a result of solving the problem, the learner has built a plan to achieve the goal and can store this plan as a set of linked actions, or as a single compound action.

The next time this type of problem is seen the whole solution can be retrieved, the given facts bound to the inputs of the stored solution, and the actions executed forward from the inputs to the final goal. The compound plan is a schema that can be used as single operator, or that can be expanded to show its internal structure. The plan is created backward from the goal, then the plan schema is retrieved and applied to show forward schema expansion.

Schema Structure

A schema defines a set of slots or variables that are linked together in a coherent structure. It defines a chunk of knowledge that can be used as a single unit in planning, or expanded to show its internal structure. Schemas can be linked to each other by sharing a slot, so a large structure can be built from smaller pieces. This large structure can in turn be wrapped in a schema that provides a set of external slots and can be used as a single unit in planning. Schemas exist at many levels of abstraction or encapsulation, and can be used at many levels of planning and design.

Code schemas

The statements in a programming language define the basic or primitive actions that are used to build a program. A statement has a plan structure that consists of an operator and its set of variables or slots, and a schema structure that defines a serial order to write these elements.

In a functional language such as Lisp, the *read* operator is a function that returns a value; this operator is shown in the top row of Figure 1 as a single action at left and as program code at right. A procedural language, such as Pascal, defines a read operator that takes a variable and sets the value of that variable. This read operator has a single slot, the variable to set; it is shown as an abstract operator in the middle left of the figure. It is also shown as the fixed and variable code that has a single slot (shown as *var*), and as a line of solution code where the slot has been filled with a specific variable.

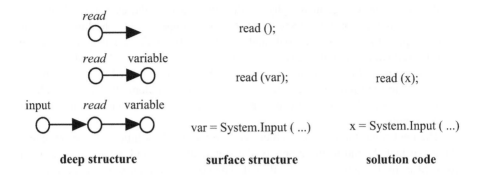

Figure 1: The read operator shown with zero, one, and two slots: as an abstract action, as empty code, and as code with its slots filled.

Some languages, such as Java, also require the user to link the action to the input stream. This form of the read operator is shown at the bottom of the figure as an operator that takes an input from a stream of values and assigns it to a variable. This language choice creates the need for a complex plan to declare and create Java input objects, that must be coded and run before the read operator can be executed. If a graphical user interface is used, then the simple act of reading in a value requires a complex plan to create, define, and display a set of input fields, labels, and buttons, and to join these together using exception processing.

Basic plan schemas
A basic or unit plan schema has its slots filled by actions, rather than by other schemas. The simplest plans are those to get an input from, and to show an output to, a user. The focus of the *read* plan is the action to read a value from the user. This is only the first step in building the plan, however, Before a user can enter the desired value, the system has tell the user what to do by showing a prompt on the screen; a *prompt* is a write action used in a read plan. The prompt action shows a string on the screen, so it takes the string to display as single input argument and, in some languages, also requires the output stream that receives the value.

The read plan is shown in Figure 2 in three forms. It is shown at top left as a single unit with one slot, the variable to set. The internal structure is shown at bottom left, where the up arrow shows the serial order between the two actions; the write must be executed before the read. The right arrows indicate data flow into and out of each operator. In some languages this plan must be coded as two separate actions, shown at top right in the figure, and in other languages the input operator can take another argument as shown at bottom right.

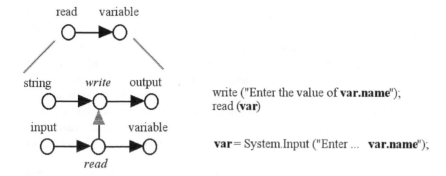

write ("Enter the value of **var.name**");
read **(var)**

var= System.Input ("Enter ...   **var.name**");

Figure 2: The read plan shown as a schema structure at left, and as two surface structures with slots at right.

The focus of a plan is the action that directly achieves the goal, and the rest of the plan is added to support the focus. The plan is built from the focus, and stored as a plan schema. When the schema is retrieved as a unit and an actual variable bound to its single slot then the slots in the read and the write actions (bottom left) can be filled automatically.

The read loop plan
A read loop reads input from a stream of values until the end of the stream is reached; if the stream of values are input, then the read loop plan can use the read plan as one element. The focus of the read loop is the read action that is controlled by a repeat statement. The structure of these basic actions is shown in two forms at the top of Figure 3. At top left is the basic control flow shown as a shaded arrow, where the read is controlled by the repeat statement. The empty code structure is shown at top right, with the single slot in bold face.

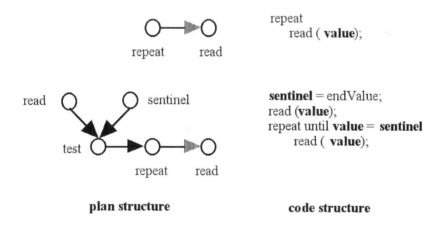

repeat
       read (**value**);

sentinel = endValue;
read **(value)**;
repeat until **value** = **sentinel**
       read (**value**);

plan structure   **code structure**

Figure 3: The read loop plan at two stages of development, shown as the plan structure and the code with slots.

The complete read loop plan schema is shown in the bottom row of Figure 3. A repeat statement has two slots, a set of actions to control and a test that stops the loop. For a read loop, the standard pattern is to define a sentinel or end value that marks the end of input, and to stop the loop when the end of the input is reached. The repeat statement controls one or more actions (control flow is shown by a shaded arrow) and uses the value of the test (data flow is shown as a solid arrow). The test uses two values, one value read in from the stream and the sentinel value. A value therefore has to be read in before the loop as shown in the serial order of the actions in the middle of the figure, and the value of the sentinel has to be set before the test is executed.

A choice has been made in this solution to set the sentinel value first and then execute the initial read, but the order of these two actions could be swapped and the plan would still be correct. Both values are used by the test, but no single linear order is imposed by this data flow. Data flow only defines a partial order on the plan actions; the actions may be implemented in many serial orders to define a correct solution.

Control flow comes in two forms that define the execution order of the actions. One line of code can control another, as occurs when an *if* or *repeat* statement controls its actions, or when one routine calls another routine. The second form of control flow is serial order, where one action is executed after another. Most plans use only data and control flow links.

The count and sum plans

The *count* and *sum* plan schemas both follow the same abstract pattern of repeating an action until the end of input occurs. The count plan adds one to a counter, and the sum plan adds a value to a running total each time the loop is executed. Each plan schema is shown in three forms in Figure 4, first the count plan (top row) and then the sum plan (bottom row).

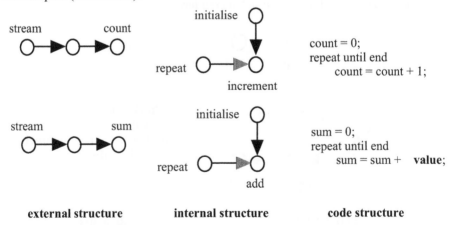

Figure 4: The count and sum plans shown as a single unit, as a plan structure and as a serial code structure.

The left column of the figure shows the external appearance or behavior of these plans. Both plans take a stream of values as input. The count plan produces the count of the values in the stream, and the sum plan produces the sum of the values in the stream. The middle column expands each plan to show the internal plan structure stored in the schema. The focus of the count plan is to increment the counter each time, and the focus of the sum plan is to add a value each time. Both plans wrap a loop, often a read loop, around the basic action to produce the stream of values to count and to sum. The right column shows the code for each plan schema placed in linear order. The variables called *count* and *sum* may be replaced by other variables with a more suitable name from the problem domain when the schema is applied.

Schema Merging

One schema may use a value produced by a second schema, so plan schemas may be linked to form a complex plan. That complex plan in turn can be stored as a schema and used as a single unit in planning by ignoring its internal structure. Schemas are merged or combined on the basis of their plan structures and not their linear structures, so the actions in a single plan may be located far apart in the final serial order of the program code.

The plan structure of a coded solution can be re-created by slicing the code backward from the goal (Horwitz, Prins, & Reps, 1989) to find the essential plan dependencies between lines of code. A simple algorithm to identify this structure is given by Rist (1989):

1. Start at the goal; the goal is the final output or result of the program.
2. Trace back through the data flow; recur.
3. Trace back through the control flow; recur.

A Pascal program to find the average rainfall for a period is shown in Figure 5, where the end of the period is signaled by a special input value of -999. The goal of the program is to find the average rainfall so the final line of the program, line 19, displays this value.

```
1.     program Noah (input, output);
2.     const endValue = -999;
3.     var  sum, rain, day: integer;
4.          average: real;
5.     begin
6.          sum := 0;
7.          day := 0;
8.          write ('Enter rainfall for day ', (day + 1), ': ');
9.          readln (rain);
10.         while rain <> endValue do begin
11.              sum := sum + rain;
12.              day := day + 1;
13.              write ('Enter rainfall for day ', (day + 1), ': ');
14.              readln (rain);
15.              end;
16.         average := 0.0;
17.         if day > 0
18.         then average := sum / day;
19.         writeln ('The average rainfall was ', average);
20. end.
```

Figure 5: A Pascal program to find the average rainfall in a period

Slicing this code traces the data and control flow back from the output at line 19. The plan structure is shown in Figure 6, where a light line indicates a data flow and a heavy line indicates a control flow. The code that shows the average is placed at the bottom of the figure. The average is set by two lines of code (lines 16 and 18), so there are two data flows into line 19. The first data flow is found at line 18, the *average* calculation. That line is now the current line; it uses two values (*day* and *sum*) and is controlled by the guard for no input.

A data flow link is traced before a control flow link, so slicing back from line 18 traces out the plan structure first for the data flow and then for the control flow. The data flow is traced backward first for the sum plan and then for the count plan. The sum plan uses an input value, so the plan structure of the read loop plan is found and traced out. At some point this backward trace reaches the terminals in the plan structure and returns back up the plan structure. After the sum, read loop, and count plans have been traced from line 18, the guard on the average calculation is added to the plan and tracing returns to line 19. The alternate data flow into line 19 is then found at line16 and added as the last step in slicing this code.

Backward slicing automatically identifies the plans in a solution. A plan is a branch of the plan structure so slicing finds and groups the actions that define each plan schema in the merged plan. The read-loop plan is shown as a group of linked plan actions at the top of Figure 6, the sum plan is shown at middle left, the count plan at middle right, and the guarded average plan appears at the bottom of the figure. A bold line number at the left of Figure 5 shows the focus of each plan.

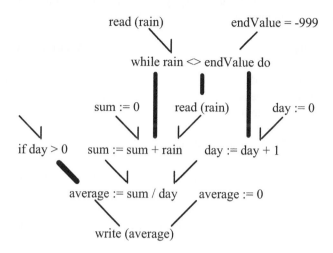

Figure 6: The plan structure of the rainfall program defined by data flow (light line) and by control flow (heavy line)

The whole *average* plan can now be stored as a single operator that takes a stream of input values and produces the average value of the elements in that stream. The internal structure of the plan is created from four smaller plans, where the plan actions are merged on the basis of their plan structure rather than the serial order of the actions in each plan. The lines of code in each plan do not appear as a single unit in the serial code order of the solution; the elements of the sum plan, for example, appear as lines 6, 12, and 18 in the code. The actions in a plan may appear as a serial chunk in the code, but are often de-localised and appear far apart in the final program structure.

Schema Creation and Retrieval

A plan schema is created backward from the goal and bottom-up from the focus, where the focus is the action that directly achieves the goal. When asked to develop a plan to count the people in a group, for example, a non-programmer starts with a verbal description of the form "add one each time you see a new person" (Bonar & Soloway, 1985). The focus of the plan is to add one each time, and this is wrapped in a loop to repeat the action as needed.

One protocol study (Rist, 1989) shows the creation of the sum plan one element at a time. The verbal protocol is shown at the left of Figure 7, and the code is shown at right in the order it appeared during the protocol. The goal of the plan is to produce the sum of the rainfall over a period. The basic operator is the plus operator, and the basic action is to add a new input value. The key insight is that the sum gets data flow from itself, because the input is added to the sum of values so far and the result stored as the new value of the sum. The complete action is a single statement and the complete plan is created bottom-up by wrapping a loop around the action, and backward by setting the initial value of the sum to zero.

Verbal statements	Written code
Rain is going to equal ... I need ...	+ rain
I need something there that keeps ...	rainfall =
I'm setting sum equal to the amount	
of rain from before, OK ...	sum
sum will be a problem, I'll come back to it later.	
[other goals solved]	
OK, do sum gets sum plus rain, or even better	rainsum = rainsum + rain;
... and now set that to zero."	rainsum = 0;

Figure 7: Backward creation of the count plan from the focus, shown in a verbal protocol and a code protocol.

Each step in plan creation is a problem for the rank novice who has to build the plan from a set of operators and variables in a new language, so every step may be visible in the design protocol. After the plan has been created once, the code for each line can be retrieved and the next design protocol might show forward expression within a line of code and backward design between lines of code. A line is coded as a single chunk, so the elements in each line appear in order from left to right. After the plan has been used several times, it can be stored as a plan schema that captures both the plan structure (data and control flow) and the surface structure (serial order and program code) of the plan. A plan schema is retrieved and expanded in the stored serial or schema order, so the next time this plan is used the code would show top-down and forward development.

A known solution shows top-down expansion through the serial order, and a new solution shows bottom-up creation through the plan structure (Rist, 1989, 1991). The overall behavior seen in software design gradually shifts from bottom-up plan creation to top-down plan retrieval as the knowledge of the designer develops and more schemas are discovered and stored. A single protocol may show a complex pattern as a schema is retrieved and expanded in a top-down pattern, and slots in the schema require new plans that are created by backward and bottom-up design.

A novice adds one basic plan at a time to an emerging solution. In the rainfall program, for example, code appears in four chunks that match the four plan schemas. A novice tries to write code in serial order from the first line of code to the last, so the first code to appear is the input. The read loop plan is coded (lines 8, 9, 10, 13, 14), then the sum plan (lines 6, 11) and then the count plan (lines 7, 12). The sum and count are then used to find the average (lines 16, 17, 18) and this average is output (line 19). When the plan schema is retrieved, it is expanded in serial order and the solution appears with no apparent thought from start to end as a single chunk of code.

A design protocol may show none, some, or all of this detailed process of software design. If code is written while the plan is being created, then the code shows a pattern of backward and bottom-up development. A good student may be able to create a plan schema backward mentally, and then code the schema as a single chunk. If there is a verbal protocol of this process, the verbal protocol would show bottom-up design and the code protocol top-down design. If there is no verbal protocol, then the code for a new plan may first appear in forward serial order.

The shift from verbal protocol to code protocol for one programmer shows an increase in forward expression of the plan actions. A series of code protocols for the same subject shows a shift from bottom-up plan creation to top-down plan retrieval (Rist, 1989). Comparison of a group of novice code protocols with a group of expert code protocols shows the same shift from bottom-up to top-down design (Rist, 1991). The change in behavior can be modeled by a system that stores a schema with both a plan and a schema structure, and shifts from the plan to the schema structure with expertise (Rist, 1995). The change in knowledge from schema creation to schema retrieval creates a new pattern of behavior that replaces backward traversal through the plan structure with forward traversal through the surface structure.

Schema Application

A schema is applied to solve a problem by filling its slots with values and variables from the problem, or with other schemas. A schema is applied in three ways: variable replacement, role filling, and plan merging.

A solution can be applied to a new problem by pattern matching, where a variable in the old solution is systematically replaced by a new variable, with no use or even knowledge of the plan structure. At a slightly higher level of abstraction, a schema can be applied by filling each of its slots with a new variable for the new problem, again with no use of the plan structure. If a change has to be made to the plan, however, replacing variable names is not sufficient and the novice cannot do the task (Kessler & Anderson, 1986). Schemas are extended and merged using the plan structure rather than the serial order of the actions.

An action in a schema may be described by an abstract role, such as initialize, validate, calculate, or use. A role describes a step in a plan; it describes a small part of a plan. A design approach based on roles uses each role to develop one part of the program at a time, in isolation from the rest of the code. Each role in turn is used to search the plan structure, and to gather together the same roles within each plan.

A role often corresponds to a type of syntactic statement. There may also be one or more plans to implement a role, such as a plan to validate the value of an input variable, so roles provide a way to store programming knowledge as well as to describe the structure of a solution (Spohrer, Soloway & Pope, 1985). The various ways to implement a role create a goal and plan or GAP tree, in which a role (called a goal by the authors) is used to retrieve a solution for each part of the problem.

Consider a plan to find the difference in seconds between two clock times, where a clock time has the form hours: minutes: seconds. The input to the plan is two clock times such as 10:24:52 and 02:45:13 and the output is an integer, the elapsed time in seconds between the input clock times. Six integers are input for the two times, and each input must be validated; all integers must be positive and less than 12 (hours), 60 (minutes) and 60 (seconds) for the parts of each time. A solution based on role structure first does all the input, then all the validation, and then all the calculation. A solution could also be organized by plan, where each value is input and validated separately, and then all the calculations are done.

The third form of schema application takes the whole plan as a unit in design, and uses the plan or schema structure to create the actions that appear in the final program code. The schema structure provides a set of roles to be filled in a given order, so it can be used to support top-down design with local slot filling. The plan structure allows plan merging through the deep structure, so it provides the most flexible and powerful approach but places a high load on memory during the design process.

Design Rules

A plan structure is converted into a solution by adding four main kinds of detail: code, order, group, and merge. Each action must be coded in some way. The actions can be placed in many linear orders. The actions can be grouped or organized into blocks in many ways, and wrapped in a routine. Finally, the actions can be merged together into a single statement, or kept separate in the code. The construction of a solution occurs in a series of moves (Schön, 1983) where a move adds structure or detail to extend the existing solution. When there is no choice, a new part of the solution is simply added at the appropriate place. When there is a choice about how to add a new piece to the solution, a design rule is used to make that choice.

Design rules appear at many levels of detail. The most basic level is the set of layout rules or conventions that define how the code looks on the page, and the naming rules that determine the names of variables. Above that level is a set of rules that guide the choice of how to code a simple deep structure, such as the use of constants rather than literals, the use of a logical expression rather than a nested selection, or the choice of a type of loop to wrap around a basic action. The design rules at the next level are very abstract and deal with either the efficiency of the code, or with modularity described in terms of coupling and cohesion (Constantine & Yourdon, 1979). Software should be designed with high cohesion (many conceptual links within a module) and low coupling (few links between modules). This is a crucial design goal, but it is difficult to give a set of operational rules that lead to high cohesion and low coupling.

Software design can be modeled by defining the knowledge of the designer as a set of plan schemas, and by defining the rules used by a designer to select one solution from the many possible variants. This section shows how each set of design rules gives rise to a different solution structure by making a different choice at a series of choice points. The deep structure of a simple problem is described first, and then the surface structure is added to create three good solutions in a functional, procedural, and object-oriented style.

Plan structure

The data flow in a plan to find the volume of a box is shown in Figure 8 at two levels of encapsulation. The plan takes three values as arguments (the length, height, and depth of the box) and produces a single value (the volume of the box) by multiplying the three dimensions. This is shown as a single chunk at the left of the figure.

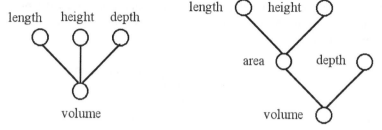

length   height   depth

length   height

area   depth

volume   volume

Figure 8: Data flow in the plan to find the volume of a box, as a single chunk (left) with internal structure (right)

The internal plan structure is shown at the right of the figure, created by decomposing the object one dimension at a time. A box is decomposed into a rectangle and a line, so the volume of a box is found by multiplying the area of the rectangle by the size of the line (the depth). A rectangle is decomposed into two lines, so the area is found by multiplying the sizes of each line (the length and height).

Procedural design rules

A procedure sets one or more values. A solution based on procedural design uses a set of assignment statements or procedures to store values in variables, and groups code statements into serial chunks on the basis of their role or syntax. Three procedural solutions to the volume problem are shown in Figure 9. The data flow is implemented by a set of assignment statements, and the code is grouped into serial chunks on the basis of three roles: input, calculation, and output.

```
read (length);                    read (length, height, depth);
read (height);                    volume = length * height * depth;
read (depth);                     write (volume);
area = length * height;
volume = area * depth;            read (length, height, depth);
write (volume);                   write (length * height * depth);
```

Figure 9: Three surface structures based on role structure (input, calculate, ouptut)

The three solutions show various amounts of merging. The solution at left follows the design rule that each action has a separate statement. The solution at top right uses a design rule that merges the same type of action into a composite statement. The solution at bottom right uses the design rule that small code is best, so the calculation occurs in the output statement.

Functional design rules

A function calculates and returns a single value, and changes nothing. A solution based on functional design uses functions to calculate values, and uses the plan structure to group actions in the code. Two functional solutions to the volume problem are shown in Figure 10.

read (length);	read (length, height, depth);
read (height);	write (volume);
area = length * height;	volume: Real **is**
read (depth);	area * depth;
volume = area * depth;	area: Real **is**
write (volume);	length * height;

Figure 10: Two surface structures based on plan structure (area, volume)

The solution at left in Figure 10 uses the same statements as the first solution in Figure 9, but groups them together on the basis of data flow rather than roles. The area calculation defines one piece of the plan structure, so the two inputs and the calculation for the area are placed together as a serial code chunk; this is a procedural solution with a plan-based serial order. The solution shown at right in the figure implements the data flow as two functions. The value returned by the *volume* function is displayed in the write statement. Merging the two functions together creates a third solution that minimizes the number of statements.

Object-oriented design rules

The basic philosophy of the object-oriented (OO) approach is to hide the actions in a set of routines, hide the data in a set of classes, and place the code in the same class as its data. A set of variables (attributes) are placed in a class and hidden by making them private. The routines (methods) that use or set those variables are placed in the same class and some of these are exported or made public. A method does a single thing, so complex methods call a set of small methods. This approach guarantees that the code is cohesive, because all the code for a single class is placed together in that class. It guarantees that code has low coupling, because all communication between classes has to pass through a set of defined method interfaces.

A system structure diagram (Rist, 1996; Rist & Terwilliger, 1995) shows the control structure of an OO system. Data flow between methods (arguments) is shown as a right arrow, and any value returned from a function is shown as a left arrow. Input within a method is shown as a down arrow pointing into the method, and output is shown as a down arrow pointing out from the method. The structure of an OO system to find the volume of a box is shown in Figure 11, where each class name is shown in bold font, the class attributes are written below the name of the class, and the class methods are shown below the attributes.

A query (attribute or function) can be placed in its correct class by using the *of* rule; here, we need to find the volume *of* a box and the area *of* a rectangle. These values are calculated so they must be functions. The *area* is a function in class *Rectangle*, calculated from the *length* and the *height* of the rectangle, so these two dimensions must be attributes in class *Rectangle*; the same reasoning places the *depth* as an attribute of class *Box*. The functions shown at bottom right of the figure define the basic structure of the solution.

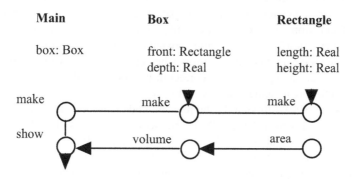

Figure 11: A solution for the box volume problem shown as a calling or system structure diagram

An output is a procedure, so a separate node is needed to output the value returned by the *volume* function in class *Box*; this has been placed in a client class that is called *Main* by convention. An input is a procedure so the values of the three dimensions must be read by a set of procedures, placed in the same class as the data. The solution structure shown in Figure 11 is completely determined by the plan structure shown in Figure 8 and the design rules described in this section. This solution structure can be stored in memory and retrieved to provide a stored answer to a known problem.

A set of design rules produces a solution from a plan structure by making a choice at each choice point in the design process. The three types of solution in this section (procedural, functional, and OO) show very different surface structures, but have the same deep structure. OO design is not different, it is more. It uses the same process of planning: plan creation from a focus through the plan structure, followed by schema retrieval and application through the surface structure. It uses the same basic elements to build software: variables and values, functions, and procedures connected by data and control flow. Object-oriented programming simply adds an additional form of grouping, class encapsulation, and adds a new set of design rules to use the power provided by class and method encapsulation.

Debugging

Planning can go wrong at any stage, creating an error or bug in the solution. Debugging a piece of software uses the same mechanism as software design - plan creation, plan merging, schema retrieval and slot filling - but the problem is not given in the specification, it is hidden somewhere in the program code. The main task is to find the bug, and then fixing it is just the normal story of adding new plans and code, or changing the existing code to support correct plan execution.

Most bugs are created not as a result of wrong knowledge, but as a result of memory overload (Anderson & Jeffries, 1985). All programmers possess the basic knowledge of programming language syntax and mechanism after their first programming course, but they continue to write buggy code all the way through their development from novice to expert. Expert design shows far fewer bugs, because an expert simply retrieves and expands a known solution with little planning, but even the expert can make errors when creating a new plan and merging it with existing code. Each plan may be correct in isolation, but the plans have to be merged in the right way to achieve the overall goal.

Program slicing is used to identify the relevant code that must be changed when a bug has to be fixed (Weiser, 1982, 1984). The incorrect output is identified and the plan that created that output is traced back from the output. If the error cannot be found by slicing, then dicing (Weiser & Lyle, 1986) can be applied. Slicing finds all the code that is used to produce the buggy output, and then the plan is divided into correct and suspect sections. Correct code is identified by slicing backward from variables that are known to be correct, and this code is ruled out as the source of the bug. The code is diced into smaller and smaller units, so the location of the bug is narrowed until the error is finally located.

People read software in order to understand it in two main ways, that may be called systematic and as-needed reading (Letovsky, 1986; Letovsky & Soloway, 1986). Systematic reading starts at the first line of code and follows the serial or linear order until the end of the code is reached and the software has been understood. As-needed reading starts at some selected location in the code, such as a variable that has to be set, and explores the software to find the other sections of code that are relevant to that variable, by tracing out the plan dependencies backwards and forwards from the initially selected code.

The location of a bug may also be found by scattering output statements through the code, but this is a brute force approach. Good problem solvers try to build an explanation by inferring a mechanism and then testing it (Chi, 1996, 2000), where poor problem solvers tend to simply change the solution and see if the new version works. This behavior may be explained as the difference between theorists and experimenters (Klahr & Dunbar, 1988). An experimenter changes the solution to see what new version does, and then tries to infer an explanation from the observed behavior. This a brute force approach, because most of the experiments do not compare rival explanations and therefore do not narrow down the search space of possible explanations. A theorist builds a theory or explanation about what is going on, and then designs an experiment to test that theory. A theorist can use a few carefully designed tests to locate the error, where an experimenter will run a long series of tests until the error is found or exhaustion sets in.

Experts do not seem to make better first guesses at the location of the bug, nor do they make more guesses. The crucial factor in improved debugging performance seems to be that experts are not as committed to any one particular explanation as novices (Vessey, 1985). They are thus able to change their theories and shift their attention as new information is gathered, where novices may discount data that does not support their theory.

Planning

The process of design creates a solution from the words in a problem specification and the knowledge in a designer's head. It maps a problem onto a solution in a series of steps, where each step adds structure to the previous representation: problem, plan, schema, code, and execution structure. The problem specification is read to identify the goals, objects, and facts in the problem. The facts are then joined to the goals through a plan structure at some level of abstraction to define a deep structure. Plan schemas are then used to build a solution by filling the schema slots and expanding each schema. At some point the code is written and executed.

The development of schemas at many levels of encapsulation and abstraction creates three patterns of behavior that characterize expert design. There is an increase in prior planning that sketches out a solution before any code is written. There is an increase in top-down design, where an initial abstract solution is progressively made more detailed. There is an increase in the evaluation of a solution both at the code, at the schema, and at the plan level. These changes may be seen as the gradual development of rules about how to design, but they are predicated on the ability to store, retrieve, and manipulate abstract schemas.

Prior planning

The code is the only solution representation for a novice, so the novice reads a problem and starts to write code in serial or execution order. Roles provides the basic schema for design, so first the inputs are coded, then the calculations, and then the outputs. The novice moves from the solution back to the problem statement during design (Green & Gilhooly, 1992; Klein & Hoffman, 1993); the code is used to link parts of the problem, so problem understanding is driven by the act of building a solution.

A professional designer plans out a solution before any code is written. The problem statement is read and reviewed to identify the goals, objects, and facts in the statement and these are linked together to define a coherent structure, the problem structure. The structure is often written down in an abstract design notation that stores enough detail to support planning, but no more (Anzai, 1991; Chi, Glaser & Rees, 1982; Petre & Blackwell, 1997). The structure of the problem is built and analyzed to find the key aspects or issues (Voss, Greene, Post & Penner, 1983) and then a solution is sketched out around these key ideas.

The truly expert designer shows very little evidence of problem solving, because the plans are simply recognized and expanded as needed. The expert shows a pattern of telegraphic or condensed speech (John-Steiner, 1985) that identifies the key decisions in the solution while leaving the rest of the solution implicit. Attention shifts instead to the highest level of planning, meta-planning (Sonnentag, 1998); an expert first works out how to approach a problem, and then the automated planning and plan knowledge produces a solution with little apparent effort.

Top-down design

Top-down design builds a solution by stepwise refinement, where detail is progressively added to an initial abstract solution until the concrete level of program code is reached. An expert has built up a rich and integrated set of schemas to store common solutions, so each slot in an abstract schema can be selected in turn and filled with a slightly less abstract schema. There are three forms of top-down design: functional, procedural, and object-oriented design.

Functional top-down design starts at the goal and retrieves a plan to achieve that goal. This plan may be concrete, such as the plan to find an area, or it may contain internal structure, such as the plan to find the volume or the plan to produce an average value from a stream of values. A complex plan encapsulates and decomposes into a set of smaller plans, so it may be described as more abstract than the contained plans but it is still concrete in the sense that it can be used as a single operator during planning. Backward and bottom-up design through the plan structure has been described in the previous sections.

Procedural top-down design
Procedural top-down design starts with an abstract solution structure and fills the slots in that solution in schema order. An expert shows a gradual increase in detail during the design process, from high-level routines to low-level code (Adelson & Soloway, 1985). If too much detail occurs at any one level, the expert groups together the mass of detail into units at a higher level of abstraction (Jeffries, Turner, Polson, & Atwood, 1981). This problem cannot be caused by simple schema expansion, because a schema has a fixed set of slots that define the complexity of the solution at the next level of detail. It may be understood as an attempt to weave together a solution from separate plans, where the detailed procedural structure only appears at the end of the plan merging process.

Pure top-down procedural design can lead to premature commitment, where the structure of the solution is frozen before any of the details are known. If there are no surprises, then this approach works fine. It is often the case, however, that one small part of the solution contains the basic idea or insight that creates the structure of the solution; this key idea has been called the focus of the solution in this paper. Novice top-down designers simply assume that each module can be defined in isolation, and may have to change the solution completely when the key problem or decision is finally encountered. Experts first identify and explore the key issues, and then build a solution around them either statically by schema retrieval and expansion or dynamically by plan creation and merging.

An expert software designer working in a new domain shows both novice and expert behavior (Adelson & Soloway, 1985). For a problem in a new domain, the structure of the problem is not known initially and has to be explored and constructed. The non-domain expert may have to write detailed code for small sections of the problem, to check that a proposed solution actually does work correctly; this is called a spike in Extreme Programming (Jeffries, Anderson & Hendrickson, 2001). Once the solutions specific to the new domain have been constructed, the expert can fit these new pieces into a large set of abstract plan schemas with little further effort.

An expert working in a new domain or on a difficult problem may work on different parts of the solution at different levels of detail. This pattern of behavior is called opportunistic design (Green, 1980; Hayes-Roth & Hayes-Roth, 1979) or serendipitous design (Guindon, Krasner, & Curtis, 1987). The designer builds outward from a set of initial ideas and alternates between different levels of description as the solution evolves and decisions made in one part of the solution affect previous decisions and decompositions.

Object-oriented top-down design

Object-oriented design includes both functional and procedural design, as well as class design. Top-down OO design specifies first the classes, then the attributes and methods in the classes, and finally the code within the methods; a method may be a function or a procedure. Both novice and expert OO programmers follow this development order (Détienne, 1995; Pennington, Lee & Rehder, 1995). Novices write an initial set of class and method names that seem reasonable, but may have no relation to the final solution. Expert OO designers use the structure of the problem to find the classes and methods needed to achieve the problem goals.

Object decomposition is reflected in a left to right order in the system structure. A use case (Jacobson, Christersson, Jonsson, & Overgaard, 1992) can be equated with a plan to achieve a goal, so the use case or plan stretches from the top-level object at the far left of the system structure to the basic at the far right. Both novice and expert OO designers generate OO code from left to right within a single goal, using top-down design or by translating the actions in the use case scenario into methods connected by control flow. The single most common transition is from a calling method to a called method (Détienne, 1995).

For the first goal or use case to be designed, this creates a pattern where a method, the attributes it uses, and the class that contains it are defined and then this three-step chunk is repeated for the next, called method (Pennington, Lee & Rehder, 1995). For subsequent goals, only the method and any new attributes are added as code is generated from left to right. The most common pattern of OO code generation can be described as goal posting, followed by goal expansion (Rist, 1996). The goals are identified from the problem statement and placed in a rough linear order; this defines a single vertical line of nodes at the far left of the system structure. Each goal in turn is then expanded to build the solution structure from left to right within a use case.

The difference between novice and expert OO design in this account is that the experts explore the problem first to identify the kernel or focus of each goal and, presumably, build the deep structure of the solution. Each plan is then coded from left to right, with the key difference that the experts know their solution will succeed.

Solution evaluation

An expert designer constantly evaluates both the quality of a solution and the quality of the design process (Sonnentag, 1996, 1998). Experts do not generate and evaluate a large number of possible solutions; instead, they evaluate their chosen (best) solution often as it is developed (Simmons & Lunetta, 1993). Evaluation of the current solution leads to a better solution to the current problem and the storage of better solution schemas, and evaluation of the design process leads to a superior ability to design. Fast and detailed feedback is essential in the development of expertise (Ericsson & Lehmann, 1996; Kluger & DeNisi, 1996) and constant evaluation allows a learner to generate this feedback without an external trainer.

A solution can be evaluated on a stylistic or aesthetic basis, as well as on the basis of complexity or difficulty. A set of design rules allow a solution to be evaluated on stylistic grounds, where the violation of a design rule (such as "one action per statement") creates a "red flag" that leads to a cycle of re-design and re-evaluation. Evaluation of a solution assumes that there is a different way to solve the problem; if there is no choice, then there is no way to improve the solution. A rank novice knows only one way to solve a problem, so has no need for solution evaluation. As design rules are learned and applied to build good solutions, the choice points and choices in design are encountered during design, and experience is developed in evaluating and choosing between different solutions.

The process of design changes with expertise to provide more feedback early in the construction of a solution. A novice tends to write all the code for a solution, and then execute the complete solution. This often leads to new cycles of design, coding, and code execution as problems are discovered when the code is run. Experts try to find any problems much earlier in the design process by adopting two main strategies. First, they simulate each part of a solution at an abstract level as soon as possible, to check that all the plans have been included in the solution and that the plan actions have been placed in a correct execution order (Adelson & Soloway, 1985). Second, they code and execute one part of a solution at a time to make debugging easier (Steier & Kant, 1985); any bug has to be located in the small section of code just added to the solution. It is much easier to find and to fix a bug in ten lines of code, than to search through hundreds of lines of code to find the bug and then identify all the code that has to be changed to fix the bug.

Problem, Deep, and Surface Structures

Three types of structure are needed to explain the variability that is characteristic of human problem solving: problem, deep, and surface structures. The problem structure (Kant, 1985; Kant & Newell, 1984) is the complex, multi-dimensional. real-world structure that defines the problem. A problem defines a set of goals and objects, but there may be many plans for a goal and one of these is selected to define the deep structure of the solution. There may be many ways to decompose a problem, and each type of decomposition defines a different deep structure. The way a problem is decomposed can lead to solutions that vary dramatically in their complexity (Ratcliff & Siddiqi, 1985; Rist, 1990; Simon & Hayes, 1976).

This paper has described a model of learning by doing in programming that uses the deep or plan structure to build a solution or surface structure. A plan is created backward and bottom-up from the goal, stored as a plan schema, and later retrieved to show forward and top-down expansion. This change in behavior may be seen at any level of detail: at a line of code, at the actions in a basic plan, at the plans in a merged plan, and at the level of large plans that are used to build a system. A solution is evaluated with a set of design rules that prefers one choice to another during solution design, and thereby leads to the creation of different solutions and the storage of different schemas. A solution is designed at the level of abstraction supported by the knowledge of the designer, leading to design at the code level for the novice and systematic top-down design for the expert. The story of schema creation, application, and evaluation provides a detailed explanation for program construction, and for learning how to program.

References

Resnick, M. (1996). Beyond the centralized mindset. *Journal of the Learning Sciences*, 5(1), 1-22.

Adelson, B. & Soloway, E. (1985). The role of domain experience in software design. IEEE Transactions on Software Engineering, 11, 1351-1360.

Anderson, J. R., & Jeffries, R. (1985). Novice LISP errors: Undetected losses of information from working memory. *Human-Computer Interaction*, **1**, 107-132.

Anzai, Y. (1991). Learning and use of representations for physics expertise. In K. A. Ericsson & J. Smith (Eds.), *Toward a general theory of expertise: Prospects and limits*, pp 64-92. Cambridge: Cambridge University Press.

Anzai, Y. & Simon, H. A. (1979). The theory of learning by doing. *Psych. Review*, **86**, 124-140.

Bonar, J. & Soloway, E. (1985). Pre-programming knowledge: A major source of misconceptions in novice programmers. *Human-Computer Interaction*, **1**, 133-161.

Chi, M. T. H. (1996). Constructing self-explanations and scaffolded explanations in tutoring. *Applied Cognitive Psychology*, 10, 33-49.

Chi, M. T. H. (2000). Self-explaining: the dual process of generating inferences and repairing mental models. In R. Glaser (Ed.), *Advances in Instructional Psychology*, pp. 161-238. Mahwah, NJ: Lawrence Erlbaum Asssociates.

Chi, M., Glaser, R. & Rees, P. (1981). Expertise in problem solving. In R. J. Sternberg (Ed.), *Advances in the psychology of human intelligence* (Vol. 1) (pp. 7-75). Hillsdale, NJ: Lawrence Erlbaum.

Constantine, L. L. & Yourdon, E. (1979). *Structured Design*. Englewood Cliffs, NJ: Prentice-Hall.

Détienne, F. (1995). Design strategies and knowledge in object-oriented programming: Effects of expertise. *Human-Computer Interaction*, **10**, 129-170.

Dreyfus, H. L. & Dreyfus, S. E. (1986). *Mind over machine*. Oxford: Basil Blackwell.

Ericsson, K. A., & Lehmann, A. C. (1996). Expert and exceptional performance: Evidence of maximal adaptation to task constraints. *Annual Review of Psychology*, **47**, 273-305.

Green, T. R. G. (1980). Programming as a cognitive activity. In H. T. Smith & T. R. G. Green (Eds.) *Human interaction with computers*, pp 271-320. New York: Academic Press.

Greene, A. J. K. & Gilhooly, K. J. (1992). Empirical advances in expertise research. In M. T. Keane & K. J. Gilhooly (Eds.), *Advances in the psychology of thinking*, pp. 45-70. New York: Harvester Wheatsheaf.

Guindon, R., Krasner, H. & Curtis, B. (1987). Breakdowns and processes during the early activities of software design by professionals. In G. M. Olson, S. Sheppard & E. Soloway (Eds.), *Empirical studies of programmers: Second workshop* (pp. 65-82). Norwood, NJ: Ablex.

Hayes-Roth, B. & Hayes-Roth, F. (1979). A cognitive model of planning. *Cognitive Science*, **3**, 275-310.

Horwitz, S., Prins, J., & Reps, T. (1989). Integrating noninterfering versions of programs. *ACM trans. on Programming Languages and Systems*, **11**, 345-387.

Jacobson, I., Christersson, M., Jonsson, P, & Overgaard, G. (1992). *Object-oriented software engineering: A Use Case driven approach*. Reading, MA: Addison-Wesley.

Jeffroes, R., Anderson, A. & Hendrickson, C. (2001). *Extreme Programming Installed*. Boston: Addison-Wesley.

Jeffries, R., Turner, A. A., Polson, P. G. & Atwood, M. E. (1981). The processes involved in designing software. In J. R. Anderson (Ed.), *Cognitive skills and their acquisition* (pp. 285-283). Hillsdale, NJ: Erlbaum.

John-Steiner, Vera (1985). *Notebooks of the mind*. Albuquerque: University of New Mexico Press.

Kant, E. (1985). Understanding and automating algorithm design. *IEEE Transactions on Software Engineering*, **11**, 1361-1374.

Kant, E. & Newell, A. (1984). Problem solving techniques for the design of algorithms. *Information Processing and Management*, **28**, 97-118.

Kessler, C. M. & Anderson, J. R. (1986). Learning flow of control: Recursive and iterative procedures. *Human-Computer Interaction*, **2**, 135-166.

Klein, G. A. & Hoffman, R. R. (1993). Seeing the invisible: Perceptual-cognitive aspects of expertise. In M. Rabinowitz (Ed.), *Cognitive Science: Foundations of Instruction*, pp. 203-226. Hillsdale, NJ: Lawrence Erlbaum.

Kluger, A. N., & DeNisi, A. (1996). The effects of feedback interventions on performance: A historical review, a meta-analysis, and a preliminary feedback intervention theory. *Psychological Bulletin*, **119**, 254-284.

Larkin, J. H. (1981). Enriching formal knowledge: A model for learning to solve textbook physics problems. In J. R. Anderson (Ed.), *Cognitive skills and their acquisition* (pp. 311-334). Hillsdale, NJ: Erlbaum.

Letovsky, S. (1986). Cognitive processes in program comprehension. In E. Soloway & S. Iyengar, *Empirical studies of programmers* (pp. 58-79). Norwood, NJ: Ablex Publishing.

Letovsky, S. & Soloway, E. (1986). Delocalized plans and program comprehension. *IEEE Software*, **3**, 41-49.

Pennington, N., Lee, A. Y. and Rehder, B. (1996). Cognitive activities and levels of abstraction in procedural and object-oriented design. *Human-Computer Interaction*, **10**, 171-226.

Petre, M. & Blackwell, A. F. (1997). A glimpse of expert programmers' mental imagery. In S. Wiedenbeck & J. Scholtz (Eds.), *Empirical Studies of Programmers: Seventh Workshop*. New York: ACM Press.

Ratcliff, B. & Siddiqui, J. I. A. (1985). An empirical investigation into problem decomposition strategies used in program design. *International Journal of Man-Machine Studies*, **22**, 77-90.

Rist, R. S. (1989). Schema creation in programming. *Cognitive Science*, **13**, 389-414.

Rist, R. S. (1990). Variability in program design: the interaction of process with knowledge. *Int. J. Man-Machine Studies*, **33**, 305-322

Rist, R. S. (1991). Knowledge creation and retrieval in program design: a comparison of novice and experienced programmers. *Human-Computer Interaction.*, **6**, 1-46.

Rist, R. S. (1995). Program structure and design. *Cognitive Science*, **19**, 507-562.

Rist, R. S. (1996). System structure and design. *Empirical Studies of Programmers: Sixth Workshop*. W. D. Gray and D. A. Boehm-Davis (Eds.), pp. 163-194. Norwood, NJ: Ablex Publishing.

Rist, R. & Terwilliger, R. (1995). *Object-oriented programming in Eiffel*. New York: Prentice-Hall.

Schön, D. A. (1983). *The reflective practitioner*. New York: Basic Books.

Simon, H. A. & Hayes, J. R. (1976). Understanding complex task instructions. In D. Klahr (Ed.), *Cognition and Instruction* (pp. 269-285). Potomac, MD: Erlbaum.

Simmons, P. E. & Lunetta, V. N. (1983). Problem-solving behaviours during a genetics computer simulation: Beyond the expert/novice dichotomy. *Journal of Research in Science Teaching*, **30**, 153-173.

Sonnentag, S. (1996). Planning and knowledge about strategies: Their relationship to work characteristics in software design. *Behaviour and Information Technology*, **15**, 213-225.

Sonnentag, S. (1998). Expertise in professional software design: A process study. *Journal of Applied Psychology*, **83**, 703-715.

Spohrer, J.C., Soloway, E. & Pope, E. (1985). A goal/plan analysis of buggy Pascal programs. *Human-computer interaction*, **1**, 163-207.

Steier, D. M. & Kant, E. (1985). The roles of execution and analysis in algorithm design. *IEEE Transactions on Software Engineering*, **11**, 1375-1386.

Vessey, I. (1985). Expertise in debugging computer programs: A process analysis. *Int. J. Man-Machine Studies*, **23**, 459-494.

Voss, J. F., Greene, T. R., Post, T. A. & Penner, B. C. (1984). Problem solving skill in the social sciences. In G. H. Bower (Ed.), *The psychology of learning and motivation* (Volume 18) (pp. 165-213). New York: Academic Press.

Weiser, M. (1982). Programmers use slices when debugging. *Communications of the ACM,* **25**, 446-452.

Weiser, M. (1982). Programmers use slices when debugging. *Communications of the ACM,* **25**, 446-452.

Weiser, M. (1984). Program slicing. *IEEE Transactions on software engineering,* **SE-10**, 352-357.

Weiser, M. & Lyle, J. (1986). Experiments on slicing-based debugging aids. In E. Soloway and S. Iyengar (Eds.), *Empirical studies of programmers* (pp. 187-197). Norwood, NJ: Ablex.

6

Algorithm Visualization

John T. Stasko & Christopher D. Hundhausen

Introduction

Algorithms and data structures are the fundamental building blocks of computational processes and techniques. They are central not only to computer programming, but to the field of computer science as a whole. An algorithm encapsulates the set of operations that are carried out to achieve some objective or to perform some task. Data structures are the logical abstractions that allow programmers to model collections of data for more convenient manipulation in algorithms.

Because algorithms and data structures are so fundamental, they are one of the first topics studied by students of computer science. A deep understanding of computer algorithms facilitates further inquiry and study of any number of areas such as databases, graphics, networks, artificial intelligence, to name but a few. For this reason, computer science instructors believe it is crucial that computer science education be built upon a strong foundation in algorithms and data structures.

As important as they are in the study computer science, most computer algorithms are highly complex. Indeed, they are composed of many steps, often including conditional or iterative operations; they often manipulate large and complex collections of data; and they are fundamentally dynamic in nature. Furthermore, descriptions of computer algorithms are largely textual—usually just prose or a pseudo code-style language. All of these features of algorithms make them difficult for students to learn.

Data structures differ somewhat from algorithms in that they are more easily conceptualized as concrete entities. Data structures are collections of data that are organized according to particular conventions. This contrasts with algorithms, which inherently seem to be more abstract descriptions of *process*. The more entity-based characteristic of data structures leads nicely to both textual and diagrammatic/graphical representations. Educators have long used graphical illustrations as explanatory aids for data structures. For instance, arrays are depicted as rows of boxes and trees are shown as nodes connected by linking arrows. Virtually every textbook about data structures includes many figures of data structure pictures.

The field of algorithm visualization takes this idea one step further. Since computer algorithms are fundamentally sequences of operations upon data structures, showing pictures of the data structures at developing points of time during an algorithm's execution will illustrate how the algorithm works. Thus, algorithm visualizations are largely dynamic illustrations of the modifications made to data over the duration of the algorithm's process. Algorithm visualizations present both the data and the operations within an algorithm. Ultimately, the purpose of an algorithm visualization is to facilitate learning by presenting a representation that is more easily understood and internalized.

It is important to note that algorithm visualization is an area within the larger field of *Software Visualization* (Price, Baecker & Small, 1993; Stasko, Domingue, Brown & Price, 1998). While the purpose of algorithm visualization is largely pedagogical, other areas of software visualization seek to help software developers in writing, analyzing, testing, debugging and optimizing their code. Although their ultimate purposes are different, the different areas of software visualization share many techniques and concepts.

In this chapter, we present an integrative review of algorithm visualization as an area of research. Our aim is to provide researchers interested in pursuing research in this area with a glimpse of where the area has come from, where it is going, and how to contribute. We begin by reviewing the evolution of algorithm visualization systems development. We then survey the substantial body of empirical studies that evaluate algorithm visualization technology. Our focus here is not only on the studies themselves, but on the varying empirical methods they have employed to study effectiveness. Following that, we present two case studies of particular algorithm visualization systems, including their design and the empirical studies that have focused on them. These case studies aim to illustrate alternative approaches to the design and evaluation of algorithm visualization, and to highlight the tradeoffs among them. Finally, we offer an agenda for future research based on our review.

Review of Technology

The notion of using illustrations and pictures to explain computer algorithms and programs is nearly as old as computer programs themselves. While a number of noteworthy program illustration efforts were carried out in the 1960s and 1970s,

Ron Baecker at the University of Toronto initiated the first concerted research studies in the area. He developed a number of systems for visualizing computer programs (Baecker, 1973; Baecker 1975) and ultimately created a video called *Sorting Out Sorting* (Baecker, 1981; Baecker 1998b) that is largely credited as being the primary stimulus for the multitude of research work that followed in the area. For a more comprehensive description of the early history of algorithm visualization and software visualization in general, see (Baecker & Price, 1998).

Initial research in the field of algorithm visualization was dominated by efforts to build algorithm visualization software systems and to expand the capabilities and expressiveness of these systems. Brown University was an early hotbed of research in the area (Bazik et al, 1998). Brown and Sedgewick's Balsa system pioneered the use of algorithm visualization as a multimedia accompaniment to computer algorithms courses (Brown & Sedgewick, 1985; Brown, 1988). Students took courses in an electronic classroom containing computer workstations. Instructors discussing particular algorithms were able to show prepared visualizations of the algorithms, and students could interact with the visualization on the computer at their desk. Instructors even included visualizations of algorithms on exams where students were challenged to "name that algorithm" upon seeing its representation. Balsa introduced many ideas for algorithm visualization including multiple different algorithm views, the interesting event notion of activating animations, and a number of algorithm representations that would be used in subsequent systems.

Closely following Balsa was Stasko's Tango system (Stasko, 1990). Tango's main focus was on the visualization development process. It introduced models and programming tools to make the creation of algorithm visualizations easier and faster. To build an algorithm visualization, a programmer utilized four simple data types: location, image, path, and transition. Tango included an animation library focused on these data types and a large set of operations upon them. More details about Tango and its descendant systems are provided later as a case study in Section 4.

Another important early algorithm visualization system was Pavane, developed by Roman and Cox (Roman et al, 1992). Pavane pioneered the idea of using a declarative approach for specifying and creating algorithm visualizations. In this approach, a developer specifies the important data and objects in the algorithm, the graphical representations of the data and objects, and most importantly, the mapping from the data to its representation. As a program runs, the system then generates the appropriate visualization and updates it during execution by enforcing the mapping rules.

Algorithm visualization systems research dominated the field until about the mid-1990's. Systems research in the area fell primarily into two main efforts— expanding the communicative expressiveness of the visualizations and facilitating the creation of the visualizations.

System expressiveness improved through a series of developments that expanded the capabilities of algorithm visualization displays. Early bitmapped personal computers and workstations facilitated multiple concurrent views of algorithms (Brown & Sedgewick, 1985). Smooth, animated transitions helped viewers track algorithm state changes (Stasko, 1990b; Stasko, 1998a). Color views provided more expressive capability (Stasko, 1990a, Brown & Hershberger, 1992). VCR-like controls allowed viewers to step through and reverse animations (Gloor, 1998). 2D views were soon joined by 3D views (Stasko & Wehrli, 1993; Brown & Najork, 1993), and sound was added for further expressiveness (Brown & Hershberger, 1992). Improved animation concurrency facilitated the visualization of parallel and distributed programs (Stasko & Kraemer, 1993; Kraemer & Stasko, 1993).

Research into visualization specification and creation also progressed during this time period. Early systems used an interesting event model of visualization activation (Brown & Sedgewick, 1985). In this approach, the visualization developer identified key points in the algorithm to have corresponding animation sequences occur. These "interesting events" often map to the high-level operations performed by an algorithm. Early algorithm visualization creation required programming in low-level graphics libraries and thus was slow and tedious. Subsequent algorithm visualization-specific libraries and toolkits made animation development easier and faster (Stasko, 1990b). Graphical, direct manipulation specification techniques sought to make authoring even more natural (Stasko, 1991; Stasko, 1998b). In these systems, the visualization developer simply drew the objects to appear in the visualization, and manually manipulated them to illustrate how they should change during program execution. Declarative specification approaches utilized a program-to-visualization mapping that would be updated automatically by the system (Roman & Cox, 1989). For more on the different styles of specification, see (Demetrescu, Finocchi, & Stasko, 2002).

Although efforts at developing new algorithm visualization systems may have peaked in the 1990's, research continues in this area today. Recent noteworthy systems include Jeliot (Haajanen et al., 1997), which provides automatic animations to students who program with special classes that have been augmented with visual operations; Leonardo (Crescenzi & Demetrescu, 2000), another declarative system in which C programs are annotated by declarative visualization statements; and ANIMAL (Rößling & Freisleben, 2002), a system including extensive internationalization, powerful extensibility, and flexible execution.

Review of Empirical Studies

In the mid-1990's, the focus of algorithm visualization research shifted markedly. Rather than concentrating on the development of algorithm visualization technology—system paradigms, specification paradigms, types of views, and the like—researchers began to turn their attention to the pedagogical effectiveness of the technology. The evaluation of algorithm visualization technology became paramount as researchers began to question their intuitions about the utility of algorithm visualizations as learning aids. In this section, we briefly review the body of empirical studies of algorithm visualization effectiveness. We organize our review around five distinct empirical methods that researchers have employed to study effectiveness: *controlled experiments, observational studies, questionnaires and surveys, ethnographic field techniques,* and *usability studies.* We describe each method, review a sample of studies that employ it, and highlight key results. Finally, to provide researchers interested in conducting future studies with guidance, we juxtapose these methods vis-à-vis several key dimensions.

Controlled Experiments

By far the most popular empirical technique for evaluating algorithm visualization technology, *controlled experiments* aim to assert a causal relationship between factors (i.e., independent variables) and measures (i.e. dependent variables). While there are many variants on controlled experiments (see Gilmore, 1990 for a review), all of the published algorithm visualization experiments have been *between-subjects experimental comparisons*. In such comparisons, participants are first screened in an attempt to ensure that they have comparable backgrounds, experience, and abilities. Next, experiment participants are randomly assigned to one of two groups, each of which is exposed to a different combination of factors that the experimenters believe to have significant effects. Third, participant performance is measured. In order for such measurements to be made, experimenters must *operationalize* the dependent variables of interest—that is, they must express them in terms of observable and measurable phenomena. Finally, in order to determine whether any measured differences are greater than those to be expected by random chance, experimenters statistically compare the measures of the alternative groups. If statistically significant differences are detected, experimenters conclude that the factors significantly affect the measures.

To date, over 20 experimental studies of algorithm visualization effectiveness have been published. The earliest of these studies aimed to establish that algorithm visualization technology could help students learn algorithms better than conventional study methods. For example, Stasko, Badre, and Lewis (1993) had two groups of students learn about the pairing heap data structure using alternative methods. One of the groups learned about the algorithm using textual materials only, while the other group had access to an animation in addition to the textual materials. On a post-test of procedural and conceptual knowledge, no significant differences were detected, although the trend was in favor of the animation group. More details of this study are presented in Section 4.

Later experimental studies began to explore the educational impact of representational features of algorithm visualization displays. For instance, in an early dissertation study, Lawrence (1993) had groups of students interact with visualizations that depicted the QuickSort algorithm operating on various input data set sizes (9, 25, or 41 elements). In addition, Lawrence varied the representation style of the data elements (horizontal sticks, vertical sticks, or dots). No significant differences were detected with respect to accuracy on a post-test of procedural and conceptual knowledge. In a later study, Lawrence designed visualizations of the Kruskal MST algorithm that used differing highlighting (color vs. monochrome) and conceptual step labeling (text labeling vs. no labeling) schemes. Students who used the monochrome highlighting significantly outperformed students who viewed color highlighting. Likewise, students who viewed textual step labels significantly outperformed students who did not view step labels.

Observing that earlier studies found only limited evidence that representation and learning medium significantly impact learning, the most recent experimental studies have explored the educational impact of differing forms of student engagement with algorithm visualization technology. For instance, Byrne, Catrambone, and Stasko (1999) considered the impact of *interactive prediction* on learning. In two separate studies adopting an identical 2 x 2 mixed-factor design, half the participants learned an algorithm using textual materials and an interactive visualization, while the other half learned the same algorithm using textual materials exclusively. In addition, half the participants were asked to predict subsequent algorithm steps (given the current step); the other half engaged in no such prediction drill. In both studies, a significant difference was found between students who viewed the visualization and/or performed the prediction, as compared to students who did neither; however, the individual effects of visualization and prediction could not be statistically disentangled. Interestingly, in a later study by Jarc, Feldman, and Heller (2000), interactive prediction was not found to significantly impact learning outcomes; however, students who performed interactive prediction spent significantly more time-on-task.

For a comprehensive review of 24 experimental studies, including a classification and analysis with respect to underlying learning theories, see (Hundhausen, Douglas, & Stasko, 2002).

Observational Studies

Typically less rigorous than controlled experiments, *observational studies* investigate some activity of interest in an exploratory, qualitative fashion.[1] Customarily, observational studies have one of two goals. They may explore phenomena about which little is known, with the goal of generating research questions and formulating hypotheses that can guide the design of more rigorous controlled experiments. Alternatively, they may be used to explore research questions that rigorous experiments are ill-suited to address—for example, questions of *how* a process unfolds, or questions concerning the nature of social interaction, which is inherently difficult to operationalize.

Over ten observational studies of algorithm visualization technology have been published to date. While the majority of these studies have relied on informal analysis techniques to draw conclusions, some studies have employed more formal techniques. In particular, Douglas *et al.* (1995, 1996) made use of *conversation analysis* (Douglas, 1995); Price (1990), Kehoe, Stasko, & Taylor (2001), and Mulholland (1998) employed *protocol analysis* (Ericsson & Simon, 1984); and Chaabouni (1996) enlisted *interaction analysis* (Jordan & Henderson, 1994). Each of these techniques advocates a slightly different set of analytical foci, and aims for results of a slightly different flavor; see (Sanderson & Fischer, 1994) for a taste of the tradeoffs involved.

The observational studies of algorithm visualization technology have used the above techniques to address a variety of research questions, including

- How do humans conceptualize algorithms? (see, e.g., Ford, 1993; Douglas et al. 1996; Chaabouni, 1996)
- How frequently is algorithm visualization used, and what role does it play, in various tasks? (see, e.g., Price, 1990; Kehoe, Stasko, & Taylor 2001; Mulholland, 1998)
- How can empirical data be used to improve the design of algorithm visualization technology? (see, e.g., Douglas et al., 1995; Chabouni, 1996)

- How can algorithm understanding be evaluated? (see, e.g., Badre et al., 1992)

For example, Ford (1993), Douglas, Hundhausen, and McKeown (1995, 1996), and Chaabouni (1996) had participants use simple art supplies to construct their own visualizations of various algorithms and programming constructs. Whereas Ford (1993) was interested in using the visualizations to understand and diagnose student misconceptions, Douglas, Hundhausen, and McKeown (1995, 1996) and Chaabouni (1996) were interested in using the visualizations as a basis for deriving algorithm visualization languages that accord with the ways in which humans conceptualize algorithms.

In contrast, Kehoe, Stasko, and Taylor (2001) had students think aloud as they completed a homework assignment with the help of various forms of visualization, including still pictures and animations. Their observations suggested that (1) students tended to use animation to learn about procedural steps of an algorithm; (2) Animations were used both to gain an overall understanding of algorithm's procedural behavior, and to refine knowledge of a specific operation; and (3) the user interface, rather than the animation itself, can be a barrier to obtaining desired information.

As a final example, Badre et al. (1992) used an observational study as a basis for designing controlled experiments of algorithm visualization technology. In their study, they used, questionnaires, observations, and exit interviews to study how students react to, and interact with algorithm animations. They found student reactions to algorithm animations to be overwhelmingly positive. However, they also discovered that it may prove difficult to disentangle a host of individual factors that influence learning with visualizations, including academic background, spatial abilities, and prior experience.

Questionnaires and Surveys
Often used as a complementary source of data in empirical studies, *questionnaires* and *surveys* elicit written responses to a set of questions in which the researcher is interested.[2] Most frequently, surveys and questionnaires request subjective data on their respondents' preferences, opinions, and advice.

Several controlled experiments and observational studies make use of questionnaires and surveys to collect *complementary* data. Such data are used in two general capacities. First, in the controlled experiments, screening questionnaire data are used in an attempt to *control for* participant experience and ability. Second, in some studies, questionnaire data are explicitly analyzed and included in the results of the study.

In contrast, a few empirical studies consider survey and questionnaire responses as their *primary* data. In these studies, all findings are drawn directly from participant responses, which, in some cases, are statistically analyzed. For example, to better understand how professors teach and conceptualize algorithms, Badre et al. (1992) surveyed 11 computer science professors at two universities. They found that over 80 percent of them used drawings and diagrams in their teaching, and that their conceptualizations of algorithms contained multiple snapshots to illustrate the dynamic aspects of algorithms.

In a more recent and comprehensive survey of computer science professors, Naps et al. (2002) report on three different questionnaires administered in conjunction with two recent computer science education conferences. The main purpose of these surveys, in which a total of 186 professors participated, was to gauge the interest in, and actual use of, algorithm visualization technology in undergraduate computer science. In one of the surveys reported (n = 29), 97 percent of respondents indicated that they demonstrate visualizations during classroom lectures at least occasionally. Roughly two-thirds said that they make visualizations available to students outside of the classroom, while roughly half indicated that they required students to use visualizations within a closed laboratory. With respect to professors' opinions regarding visualization effectiveness, 95% of the respondents to two of the surveys agreed or strongly agreed with the statement that "using visualizations can help learners learn computing concepts." However, two thirds of the respondents of one of the surveys cited visualization development time as a major impediment to using algorithm visualization technology.

Besides professors, students have been the other major respondents to surveys of algorithm visualization effectiveness. For instance, Lawrence (1993) elicited the preferences of undergraduate computer science students with respect to algorithm visualization display styles. She later found that students' preferred display styles did not lead to better learning. Stasko (1997), in contrast, used a questionnaire to elicit students' impressions of the required algorithm animation assignments in a course they had taken. He found that that their impressions were almost universally positive, and that they believed that the animation assignments had helped them learn.

Ethnographic Field Techniques

Distinguished by their commitment to collecting data in naturalistic settings, *ethnographic field techniques* include any of the qualitative techniques one might use to conduct a field study (Sanjek, 1995).[3] Perhaps the most eminent of these techniques, *participant observation*, is predicated on the idea that one can best gain an insider's view on the setting of interest by actually participating in it as an accepted member. Instead of merely observing members of the setting as they go about their business, the participant observer gradually gains the acceptance of her informants, allowing her to take on an increasingly participatory role in the activities of the setting. Other widely-used ethnographic field techniques include interviewing[4], artifact collection, diary keeping, and fieldnotes; see (Wolcott, 1992) for a taxonomy.

At least three studies of algorithm visualization use ethnographic field techniques for complementary data, while a fourth uses them as its primary source of data. Both Badre *et al.* (1992) and Price (1990) used *exit interviews* in essentially the same way as other studies use exit questionnaires: to elicit participants' comments regarding their experiences in the study. Concurring with participants who responded to questionnaires and surveys in other studies of algorithm visualization, participants in both of these studies said in interviews that they believed that animation had enhanced their understanding of the algorithm in some way. Ford (1993), in contrast, interviewed his participants in order to gain an understanding of the visualizations they had designed during the course of his study.

In the only ethnographic study of algorithm visualization technology of which we are aware, Hundhausen (2002) used participant observation, interviewing, artifact collection, and videotape analysis to study a third year undergraduate classroom in which students constructed and presented their own visualizations of the algorithms under study. He found that, when students used computer-based algorithm visualization technology to construct their visualizations, they tended to lose their focus on the algorithms under study, and instead become steeped in low-level implementation issues. In contrast, when students used art supplies (paper, pen, scissors, etc.) to construct their visualizations, they spent far less time implementing their visualizations, and the time that they did spend was more focused on the algorithms under study. Hundhausen used these findings as a basis for developing alternative, "low fidelity" algorithm visualization software (Hundhausen & Douglas, 2002).

Usability Studies

A special kind of observational study, the *usability test*[5] aims to identify and diagnose problems with an interactive system's user interface. In a usability test, researchers give a small number of participants, who may work in pairs[6] (usually three to five individuals or pairs in total), a set of tasks to perform with the system under study. The tasks are chosen so as to engage participants in scenarios that the system's designers believe the system should be able to handle. All interaction between participants and system is captured on videotape; in addition, the researchers typically take detailed notes during participant sessions. By reviewing the videotape and their notes, researchers pinpoint *breakdowns* in human-system interaction that may indicate problems with the user interface. By determining the cause of those breakdowns, they may be able to suggest ways of changing the interface such that users will not encounter the problems in the future.

Unfortunately, usability studies of algorithm visualization technology either are not routinely conducted, or go unpublished. Indeed, we are aware of only two such usability studies. To tune the Paradocs algorithm visualization system for a controlled experiment, Price (1990) ran eight participants through a pilot usability test. In work reported in conjunction with their observational study of human visualization, Douglas, Hundhausen, and McKeown (1995) report on a usability study they conducted of the Lens system (Mukherjea & Stasko, 1994).

Juxtaposition of Methods

For the researcher interested in using empirical methods to explore algorithm visualization technology, we conclude this section by summarizing the various techniques, and by offering some guidance on how to choose an appropriate technique. Table 1 provides a synopsis of each kind of research study along four key dimensions: unit of analysis, data collection method, analysis method, and desired results. The "Desired Results" column provides a logical starting point for understanding the differences in effectiveness statements made by studies employing each technique. Controlled experiments enlist the Scientific Method in an attempt to establish a *causal link* between algorithm visualization technology and effectiveness. Generally less rigorous, the other methods aim to make broader, qualitative statements about algorithm visualization effectiveness, or about the conditions that might lead to it. Questionnaires and surveys, for instance, seek *individual preferences* or *opinions*, which may reflect people's subjective experiences with algorithm visualization technology.

In light of the range of different kinds of results for which the research techniques strive, it should come as no surprise that they recommend a corresponding range of data collection and analysis techniques. For example, in order to assert causality, controlled experiments must adhere to the stringent requirements of the experimental methodology upon which they are based. By contrast, observational studies are interested in producing qualitative accounts whose plausibility is firmly grounded in empirical data. As a result, they enlist a markedly different data collection and analysis techniques, which may be every bit as systematic as those employed by controlled experiments (see, e.g., Jordan & Henderson, 1994), but which focus on videotaped episodes instead of performance measures.

The techniques' units of analysis depend more on the *intellectual tradition* (Sanderson & Fischer, 1994) or community of practice with which they are associated than on the form of their desired results. First, the Behavioral tradition within which controlled experiments are conducted, as well as the Cognitivist tradition[7] out of which usability studies, protocol analysis-based observational studies, and analytic techniques have evolved, has a longstanding interest in *individual* cognition. As a consequence, those who conduct controlled experiments tend to operationalize and measure *individual* behavior, and those who perform analytical evaluation, usability studies, and protocol analysis-based observational studies tend to focus on *individuals* performing tasks.[8] In contrast, many observational techniques, as well as ethnographic field techniques, have evolved out of the Social tradition, whose roots lie in sociology and anthropology. As a result, their units of analysis tend to involve groups of interacting people.

Technique	Unit of analysis	Data Collection Method	Analysis method	Desired results
Controlled experiments	Individual knowledge or behavior	Measurement of predefined behavioral observables, such as test scores and test-taking time	Statistical tests	Statistically-significant differences that favor algorithm visualization technology
Observational studies	Individuals or groups engaged in SV tasks	Videotaping, interviews, questionnaires, surveys	Protocol analysis, Conversation analysis, Interaction analysis, and others	Qualitative accounts of processes by which humans engage in observed tasks, ideas for future research
Questionnaires and surveys	Individuals	Written questionnaires and surveys	Counts, catalogs, statistical tests	Quantitative or qualitative accounts of individual preferences or impressions; often used to complement other methods
Ethnographic field techniques	Recurrent social scenes	Field notes, participant observation, interviews, videotaping, artifact collection, diary keeping, and others	Interview transcription and analysis, field note consolidation, to name but a few	Qualitative descriptions of shared cultural knowledge and practices from the perspective of an insider.
Usability studies	Individuals or group performing tasks with a specific SV system	Videotaping, questionnaires, interviews	Same as for observational studies	List of usability problems, along with possible design solutions

Table 1. Alternative research methods vis-à-vis four key dimensions

That each research technique produces effectiveness statements of a different flavor implies that different techniques will be appropriate for different types of research questions. Selecting an appropriate technique thus involves finding a match between one's research questions and a technique that can provide satisfactory answers to those questions. To provide some guidance, Table 2 suggests a set of general SV effectiveness questions that each of the techniques is designed to address. For instance, the table indicates that, to identify and diagnose the problems in a prototype algorithm visualization system, one should employ a *usability test*.

Answers to the research questions listed in the second column of the table do not come without a price; associated with any research technique is a set of pragmatic and methodological tradeoffs to be considered. Columns 3 and 4 of the table list some of the most important of these for each technique. For example, while the results of controlled experiments are generally perceived as hard evidence of effectiveness, it may be difficult, from a practical standpoint, to stage such a study, as Stasko *et al.* (1993) point out:

> Pragmatically, it is challenging to assemble the appropriate ingredients for a [controlled experiment]. . .[A] group of subjects who are at an appropriate point in their educational careers must be available. Even this may not be enough, however, because splitting the subjects into two groups, one using animation and one not using animation, may unfairly influence student achievement and grading the particular course in which the students are enrolled. (p. 61)

While observational studies can provide detailed accounts of the processes by which humans use algorithm visualization technology, small sample sizes limit the extent to which those accounts can be generalized. Questionnaires and surveys provide an ideal means of collecting subjective data; however, it may be difficult to determine whether respondents provided accurate answers, or answers they believed the researcher wanted. Ethnographic techniques are well suited for obtaining an insider's perspective on the artifacts, activities, and beliefs of a particular cultural scene. To obtain such an insider's perspective, however, typically requires a relatively long time in the field—typically at least two to three months. Finally, while usability studies are effective for identifying usability problems with interactive SV systems, it may be difficult to glean solutions to those problems from a usability test. Further, their narrow scope limits them from addressing questions that lie beyond usability (e.g., usefulness).

Technique	Appropriate Research ?'s	Advantages	Disadvantages
Controlled Experiments	Does some SV technology factor cause some effect, where that effect can be precisely stated in terms of measurable and observable phenomena?	• Results are quantitative and thus seen as hard evidence • Results have high degree of objectivity, since they are based on pre-determined observable and measurable	• Studies require large sample sizes to meet the assumptions of the statistical models • Participants must be carefully screened; it may be difficult to find study

		variables • Experiments can be replicated, which lends to their credibility • If requirements of statistical models are met, results may be generalized to population	participants who meet the selection criteria. • It is difficult to control for all possible factors, and hence to assert causality • Difficult to ensure that experimental conditions have equal access to equivalent information • Controlled experimental settings reduce ecological validity
Observational Studies	How do humans engage in an SV task? What resources do they make use of, and how?	• Can provide fine-grained accounts of how humans engage in SV tasks • Studies can be conducted with a small number of participants • Can generate research questions and hypothesis for future study • Well-suited to study of collaborative tasks	• Difficult to generalize human behavior from small number of participants • Qualitative results may appear soft, may not be regarded as hard evidence • Since results rely on post-hoc analysis, objectivity may appear damaged
Questionnaires and Surveys	What are user preferences with respect to a particular visualization or visualization system? What do they wish they had had? What would they like to see in future versions?	• Can guarantee anonymity • Unintrusive, especially if administered via e-mail • Efficient means of obtaining complementary subjective data on participants' preferences and opinions in an empirical study	• Sometimes difficult to know which questions to ask • Difficult to know whether respondents provided answers that are indicative of their true opinions, or whether they provided answers they believed the researcher wanted.

Ethnographic field studies	How might SV technology fit in to the overall practices of a cultural scene? What knowledge about an SV artifact is shared by members of cultural scene?	• In the early stages of research, can help to determine ways in which SV technology might fit into the activities in a cultural scene • High degree of ecological validity	• May be difficult, from a practical standpoint, to arrange fieldwork • May be time-consuming; in typical ethnographic fieldwork, one needs to conduct a minimum of 2-3 months of fieldwork to "get into" culture • Unit of analysis too broad to provide insight into detailed human-visualization interaction
Usability Studies	What are the usability problems with an SV system, and how might we fix them?	• Effective means of evaluating the user interface of an interactive system • Three to five participants sufficient to identify most of a system's problems	• Narrow scope: Does not consider questions beyond those at the level of tasks and the user interface • Can be difficult to find design solutions to fix problems

Table 2. The advantages, disadvantages, and scope of the alternative research techniques

Case Studies

To make the reviews of the previous two sections more concrete, we now turn our attention to two case studies of our own algorithm visualization technology development and evaluation efforts. The first case study considers the evolution of Stasko's series of algorithm visualization systems: TANGO (Stasko, 1989), POLKA (Stasko & Kraemer, 1993), and, most recently, SAMBA (Stasko, 1997). The second case study focuses on ALVIS (Hundhausen, 1999; Hundhausen & Douglas, 2002), a sketch-based, "low fidelity" algorithm visualization system that differs markedly from Stasko's systems. Our goal, in presenting these case studies, is not only to illustrate the alternative ways in algorithm visualization technology has been designed, but also to consider the differing roles of empirical studies in the design and evaluation of the technology.

		• variables • Experiments can be replicated, which lends to their credibility • If requirements of statistical models are met, results may be generalized to population	participants who meet the selection criteria. • It is difficult to control for all possible factors, and hence to assert causality • Difficult to ensure that experimental conditions have equal access to equivalent information • Controlled experimental settings reduce ecological validity
Observational Studies	How do humans engage in an SV task? What resources do they make use of, and how?	• Can provide fine-grained accounts of how humans engage in SV tasks • Studies can be conducted with a small number of participants • Can generate research questions and hypothesis for future study • Well-suited to study of collaborative tasks	• Difficult to generalize human behavior from small number of participants • Qualitative results may appear soft, may not be regarded as hard evidence • Since results rely on post-hoc analysis, objectivity may appear damaged
Questionnaires and Surveys	What are user preferences with respect to a particular visualization or visualization system? What do they wish they had had? What would they like to see in future versions?	• Can guarantee anonymity • Unintrusive, especially if administered via e-mail • Efficient means of obtaining complementary subjective data on participants' preferences and opinions in an empirical study	• Sometimes difficult to know which questions to ask • Difficult to know whether respondents provided answers that are indicative of their true opinions, or whether they provided answers they believed the researcher wanted.

perform different Actions, such as movements, color changes, and size changes, at particular animation times. Thus, just by scheduling different Actions to overlap in duration, concurrent animation effects, either in the same algorithm views or across algorithm views, could be achieved. This capability was important in developing visualizations of parallel and distributed programs in which illustrations of the concurrency of operations were crucial. Polka was used extensively to build visualizations of many different kinds of parallel and distributed programs.

The code sample below illustrates a simple animation routine in Polka. In it, two rectangle objects smoothly exchange positions as might be shown in a visualization of a sorting algorithm.

```
int
Rects::Exchange(int i, int j)
{
Rectangle *temp;
Loc *loc1,*loc2;
int len;

loc1 = blocks[i]->Where(PART_SW);
loc2 = blocks[j]->Where(PART_SW);
Action a("MOVE",loc1,loc2,1);
Action *b = a.Reverse();
len = blocks[i]->Program(time,&a);
len = blocks[j]->Program(time,b);
time = Animate(time,len);

temp = blocks[i]; blocks[i] = blocks[j];
blocks[j] = temp;
return(len);
}
```

The Tango and Polka systems defined a clear boundary between the visualization developer and the visualization viewer. Although building visualizations with Tango or Polka was easier than with low-level graphics toolkits, an animation developer still had to be a competent C or C++ programmer and had to learn the algorithm visualization library. Therefore, visualization developers who learned the systems typically built animations for students and other people to watch. The need for a system that was even more accessible and could allow undergraduate students without graphics or user interface programming experience to build algorithm visualizations was evident. A more recent system, Samba, provides that capability by introducing a simple scripting language that facilitates students creating their own visualizations (Stasko, 1997). Each line of a script specifies an animation operation. The first word of a line identifies the animation command to perform. Subsequent fields on the line denote graphical objects being modified and parameters to the animation action. Samba acts simply like an animation interpreter by reading textual script files that contain animation directives and performing the specified actions. Samba is implemented using Polka–it acts as a front-end animation application. Below we list a section of an example Samba script file. A web-based version of Samba written in Java also was introduced.

```
circle 1 0.8 0.8 0.1 red half
line 2 0.1 0.1 0.2 0.2 green thin
rectangle 3 0.1 0.9 0.1 0.1 blue solid
text 4 0.0 0.0 0 black Hello
text 5 0.5 0.5 1 black
RealLongStringandThenSomeAndEvenMore
circle 6 0.3 0.3 0.2 wheat solid
triangle 7 0.5 1.0 0.6 0.8 0.4 0.9 cyan solid
bigtext 8 0.2 0.2 0 black Some Big Text
moveto 1 6
moverelative 3 0.05 -0.4
jumprelative 4 0.4 0.4
raise 1
lower 1
color 6 blue
move 3 0.5 0.5
bigtext 8 0.6 0.2 0 DeepPink More Big Text
flextext 88 0.4 0.3 0 magenta 8x13bold Flex Text
8x13bold
rectangle 12 0.7 0.7 0.1 0.1 green solid
exchangepos 12 3
switchpos 12 3
circle 99 0.8 0.8 0.15 black outline
exchangepos 1 99
bg pink
coords -0.5 -0.5 1.5 1.5
```

Samba's main use came as an instructional tool in a computer algorithms course. This course was traditionally very theoretical with homework assignments consisting of analysis problems and proofs. Samba allowed us to create a new kind of homework assignment, one in which students implemented a computer algorithm in a programming language of their choice, and then they supplemented the program with output statements to generate commands in the Samba scripting language. The students' challenge was to make the Samba commands illustrate a visualization of the algorithm as it executed. We found that the undergraduate students were able to quickly learn the Samba commands and were able to build quite interesting visualizations of the different algorithms discussed in the course.

Empirical Studies
The fundamental question that instructors care about with respect to algorithm visualization is "Do the visualizations help students learn about algorithms?" Many related issues and questions do arise, but fundamentally all are related to this primary query.

A natural instinct is to assess this question in a direct manner by comparing two different groups of students learning a new algorithm. The two groups will utilize exactly the same learning materials, except that the one group also will have an algorithm visualization to use. After an initial "learning" period, we will test the student's knowledge about the algorithm through an exam. This style of controlled experimental evaluation, the first one discussed in the Empirical Studies review section earlier, utilizes a classical, comparative experimental methodology found in the sciences and widely used by psychologists. First, we posit a hypothesis. Here, our hypothesis is that algorithm visualization will help students learn about an algorithm. Next, we manipulate an independent variable. Here, that variable is whether algorithm visualization is present or not. We then measure performance via a dependent variable. In this case, the dependent variable is the student's score on an exam evaluating his or her knowledge of the algorithm.

One past study (Stasko, Badre, & Lewis, 1993) focusing on graduate students learning a new algorithm is very typical of experiments of this type. Twenty graduate students were given the objective of learning about the pairing heap data structure and a set of operations on that data structure. (Taken together, we consider this the *pairing heap algorithm*.) The students were divided into two groups of ten. Each group was given a 45-minute session to learn about the pairing heap.

The first group's learning materials consisted of a set of pages containing a textual description of the algorithm. The description did include pictures and figures, and it described the operations on the data structure and how they are implemented. This group studied the text for the 45 minutes.

The second group utilized the same text materials for the first 30 minutes, but also was allowed to interact with a Tango visualization of the algorithm for the final 15 minutes of the learning session. No pre-set animation operations were shown. Rather, the participants could enter their own data into pairing heaps and try out any operations possible.

After the 45-minute introductory session, all learning materials were removed and the participants took a post-test evaluating their knowledge of the algorithm. Participants had a maximum 45 minutes to work on the post-test.

The exam consisted of 24 questions testing knowledge of the algorithm in a variety of ways. Different questions tested students' knowledge of invariant declarative or factual properties of the algorithm, computational complexity issues of the algorithm, and how the algorithm's operations worked procedurally.

Final results favored the group of students who interacted with the animation. They averaged 13.6 correct replies versus 11.2 correct replies for the group using only the text materials. Note however that this difference was not statistically significant using a two-sample t-test ($t=1.11$, $df=18$, $p<0.13$). Thus, we cannot safely conclude that the presence of the algorithm visualization affected learning.

This result is fairly typical of that found in other early studies in this area, and is a primary reason that we included it here. This experiment and other early ones like it illustrated that one cannot merely "throw visualization at an algorithm" and expect it to strongly facilitate students' learning. More specifically, early experiments helped identify a number of key issues to consider more deeply in evaluating the effectiveness of visualizations.

First, one must question the quality of any learning materials including visualizations. Do the materials provide the appropriate content and explanations to assist students in grasping these new algorithms? In the pairing heaps study, an informal walkthrough of the learning materials cast doubts on whether students would be able to understand the concepts being introduced. The relatively low scores on the exam seemed to confirm that view.

Next, how a learner interacts with an algorithm visualization is a crucial issue. In the pairing heap study, students were merely presented with the visualization software. They were responsible for entering trial data to observe the heap operations. It is questionable whether someone learning a new concept has the knowledge to generate appropriate test cases for the visualization. Furthermore, this particular visualization had no accompanying audio explanations, object labels, and so on, thus making it more challenging for students to interpret the visual mapping from the data structure to its visual representation. Many algorithm visualizations are created to be used in conjunction with an instructor's explanation of the visual mapping, what operations are occurring, what to watch for, etc.

Finally, identifying benefits from such a modest interaction with learning materials is simply a difficult proposition. Could students learn about this relatively complex algorithm in 45 minutes? Would the presence of a visualization for 15 minutes really affect the learning outcome? In retrospect, it seemed overly optimistic to believe that both questions would be answered in the affirmative.

The key contribution of this study and other initial ones in algorithm visualization was to force researchers to look more closely at their assumptions about the pedagogical effectiveness of the visualizations and to consider alternative ways of evaluating visualizations' learning benefits.

In a follow-on study involving an algorithm visualization from the Polka system, we utilized an experimental session that mimicked a homework situation rather than an exam situation (Kehoe, Stasko, & Taylor, 2001). In this study, graduate student participants were again divided into two groups, with one having access to an algorithm visualization, just as in the experiment previously discussed. The algorithm being studied in this experiment was the binomial heap data structure and the set of operations on it. This study differed in that no explicit learning and testing periods were used. Rather, student participants received both the learning materials and the set of questions to be answered at the beginning of a session. Participants were instructed to use the instructional materials to learn about the binomial heap and answer the exam questions. Also, sessions had no time limit–students could work for as long as desired.

Participants in the algorithm visualization group answered an average of 20.5 questions correctly out of 23 (std. dev. 2.9). Participants in the group without visualization access answered 16.0 questions correctly (std. dev. 3.2). This was a statistically significant difference (t=2.55, p<0.029), thus leading us to conclude that the presence of the algorithm visualization did aid learning. More importantly however, we carefully observed each session and noted how the students confronted the learning challenge and when the students utilized different types of learning materials. In addition, we interviewed each student in-depth at the end of a session.

The key qualitative difference noted between the two groups was that the group using the algorithm visualization was more upbeat and engaged throughout, and they generally worked longer on the problems. Interviewed afterward, all of the students in this group commented that the visualization was helpful in learning the algorithm. They felt that it contributed to their understanding of the operations on the binomial heap data structure. Conversely, students in the group without the algorithm visualization seemed to labor more during the learning sessions. They frequently commented about the difficulty of the material. After their sessions, these students were shown the algorithm visualization and asked about it. Five of the six students in this group felt that it would have helped them learn the binomial heap better, with a number making very specific comments about particular misunderstandings that became clear upon viewing the visualization.

With respect to the Samba system, no formal experimental studies were performed, but the system was used in actual algorithms courses for a number of years. Our evaluation measures of Samba consisted largely of observations, anecdotal evidence, and assessments of student opinions through questionnaires and surveys. On the whole, students seemed to enjoy the assignments and, in particular, the change of pace from the very theoretical material making up the majority of the course. Students became quite competitive with each other in terms of trying to "top" other students' animations. Finally, we observed that the animation assignments appeared to help students learn the particular algorithms being visualized. Performance on final exam questions about those algorithms was extremely good, definitely higher than on questions about other algorithms. Of course, the Samba visualization assignments facilitated the students spending much more time working on those particular algorithms, but the students seemed to enjoy doing so, and they did gain a deeper understanding from the interaction.

ALVIS

We now consider ALVIS (Hundhausen, 1999; Hundhausen & Douglas, 2002), a sketch-based algorithm visualization system that differs markedly from the three systems just discussed. We begin with an overview of the ALVIS environment, focusing on its support for two key tasks: algorithm visualization construction and algorithm visualization presentation. We then step back and consider the series of empirical studies, spanning nearly five years, out of which its design evolved. .

System Overview

ALVIS aims to make the construction of computer-based algorithm visualizations as simple as creating "homemade" visualizations out of simple art supplies. Accordingly, its conceptual model is rooted in a design session in which one uses pen, paper, scissors, and other art supplies to quickly create a homemade visualization. In particular, ALVIS users enlist a pen tool to sketch visualization objects (called "cutouts" in ALVIS) on virtual pieces of scrap paper that they have cut out using a scissors tool. They then lay out these virtual pieces of scrap paper on an animation canvas by direct manipulation. Finally, they quickly define animations by dragging out their paths (see Figure 1).

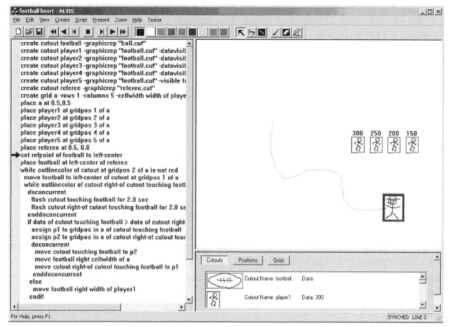

Figure 2. Example Alvis view display.

ALVIS is built upon a high-level scripting language called SALSA (Spatial Algorithmic Language for Storyboarding). When ALVIS users create animation objects, lay them out, and animate them by direct manipulation, they are actually generating SALSA commands as a byproduct. These SALSA commands appear in the "script window" on the left-hand side of the environment in response to meaningful actions, thus giving users feedback on their interface-level actions. For example, in response to choosing "Cutout…" from the "Create" menu, filling in the dialog box, and clicking OK, ALVIS inserts into the script view a corresponding SALSA "Create" command:

```
create cutout foo -graphicrep "arrow.cut" -data "11" -
datavisible true
```

Of course, if constructing a complete algorithm visualization were as simple as creating a sketch in a drawing editor, we would not need algorithm visualization technology; indeed, direct manipulation alone is not a sufficient means of defining most algorithm visualizations. The tricky part lies in animating objects such that they model the (often complex) conditional and iterative execution of the underlying target algorithm. To model algorithm execution, the SALSA language provides conventional conditional (*if-else*) and iterative (*while* and *for-each*) constructs, coupled with explicit support for *spatial relations* such as *above*, *below*, *left-of*, and *right-of*. Thus, ALVIS users can use the relationships among objects in an animation as the basis for formulating the algorithmic logic they are modeling.

For example, suppose we would like to iterate through an array of values, changing each value to "0" along the way. In ALVIS, we can model an array of values as a *grid* of *cutouts*. Using a combination of direct manipulation and dialog box fill-in, we first create and place the cutouts and the grid, generating the following SALSA code in the process:

```
create grid MyGrid -rows 1 -columns -4
place MyGrid at 0.5, 0.5
create cutout Cutout1 -graphicrep "object.cut" -data
"11" -datavisible true
create cutout Cutout2 as clone of Cutout1 -data "20"
create cutout Cutout3 as clone of Cutout1 -data "24"
create cutout Cutout4 as clone of Cutout1 -data "3"
place Cutout1 at gridpos 1 of MyGrid
place Cutout2 at gridpos 2 of MyGrid
place Cutout3 at gridpos 3 of MyGrid
place Cuout4 at gridpos 4 of MyGrid
```

To mark the item currently being considered, we next create another cutout that looks like a pointer. Initially, we place that pointer above the first item in the grid. In response to our actions in the ALVIS environment, suppose the following SALSA code is generated:

```
create cutout pointer -graphicrep "arrow.cut"
place pointer above gridpos 1 of MyGrid
```

We could then use the following while loop to iterate through all the cutouts:

```
while (pointer is-left-of right-edge of MyGrid)
 set data of cutout below pointer to "0"
 move pointer right cellwidth of MyGrid
endwhile
```

In addition to supporting algorithm visualization creation the ALVIS environment contains three features specifically designed to facilitate the interactive presentation of algorithm visualizations. First, a presenter can execute a script both forwards and *backwards*, thus allowing specific pieces of an animation to be replayed in response to audience questions. Second, since SALSA is an interpreted language, a script can be dynamically modified at the point at which its execution is currently halted. This promotes audience participation by allowing the presenter to take their suggestions into consideration on the spot, and by facilitating "what-if" analyses with alternative sets of input data. Third, ALVIS provides pen tools with which the presenter and audience members can dynamically mark up and annotate an algorithm visualization. Such markings and annotations are not part of the animation itself, and can be easily erased with an eraser tool.

Empirical Studies

The design of ALVIS has evolved out of a series of empirical studies. Below, we chronicle that evolution, beginning with our early observational studies of the human conceptualization of algorithms, and then moving to our ethnographic studies of algorithm visualization construction and presentation in an undergraduate algorithms course. Because it is similar in design to the experimental studies reported in the previous case study, we do not report an experimental evaluation of ALVIS's low fidelity algorithm visualization construction technique here (see Hundhausen & Douglas, 2000).

Observational Studies

In an unpublished usability study we conducted on the Lens visual debugging system (Mukherjea & Stasko, 1994), we discovered that the process of mapping algorithm source code to a visual representation can be difficult for a visualization system user. This proved to be especially true if the visualization primitives provided by the visualization system did not accord with the user's conceptualization of the algorithm being visualized. In response to this discovery, we decided to augment our usability study with a paper prototype study. Before usability participants (who were computer science graduate students) worked with Lens, they were asked to use simple art supplies (paper, pens, scissors, etc.) to construct their own, homemade visualizations of the algorithm (Bubblesort) they would ultimately visualize in Lens.

In our publications on this study (Douglas, Hundhausen, & McKeown, 1995, 1996), we report two key observations:

1. Participants found art supplies extremely natural and easy to use as a medium for visualization expression.
2. Participants' conceptualizations of the target bubblesort algorithm exhibited remarkable similarities at a semantic level.

The first observation constitutes the original inspiration for ALVIS's conceptual model. Indeed, we came to see art supplies as extremely natural materials for visualization construction. Further analysis of these "art supply" sessions, both in these studies, and in later studies conducted by Chaabouni (1996), enabled us to identify the importance of *spatial relations* in the construction of algorithm visualizations. For example, one visualization of the insertion sort algorithm [collected by Chaabouni (1996)] uses a special arrow to mark the dividing line between the sorted and unsorted portions of the list (see Figure 2). Each time the outer loop of the insertion sort is executed, the arrow moves one position to the right.

Notice that the internal logic governing the insertion sort *algorithm* is distinct from, but parallel to, the internal logic governing the insertion sort *animation*. For instance, consider the way in which each decides when the array is sorted. In the insertion sort *algorithm*, we continue sorting until the outer loop index is one less than the size of the array *a*. By contrast, in the insertion sort *animation*, we observed that its creators continued with the animation until the arrow slid to the right of the row of blocks. Whereas in the algorithm, the internal logic is mathematical (*xindex = n – 1*), in participants' execution of the animation, the internal logic was spatial (*arrow is right of edge of row of blocks*). We identified precisely this kind of spatial logic, which is analogous to the mathematical logic of an algorithm, as an expressiveness requirement of SALSA.

Ethnographic Studies

A few years after the original art supply studies, we became interested in learning exercises in which students construct and present their own visualizations of the algorithms they study. To investigate the educational value of such exercises, we conducted a series of ethnographic studies in consecutive offerings of an undergraduate course in which students were required to create their own visualizations of an algorithm they had studied, and to present their visualizations to their instructor and peers for feedback and discussion (Hundhausen, 2002). To study students' use of algorithm visualization technology within the context of these assignments, we employed a variety of ethnographic field techniques, including participant observation, semi-structured interviews, videotape analysis, diary collection, and artifact analysis.

In the first of our ethnographic studies, students used the SAMBA algorithm animation package (Stasko, 1997) to construct visualizations that (a) were capable of illustrating the algorithm for arbitrary input, and (b) tended to have the polished appearance and precision of textbook figures, owing to the fact that they were generated automatically as a byproduct of algorithm execution. We found that students spent 33.2 hours on average (*n* = 20) constructing and refining a single visualization in SAMBA. They spent most of that time steeped in *low-level graphics programming*—for example, writing general-purpose graphics routines capable of laying out and updating their visualizations for any reasonable set of input data.

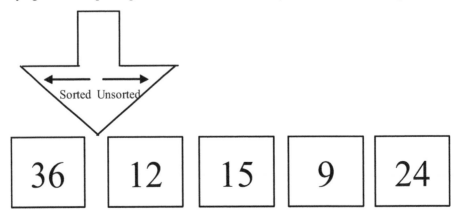

Figure 2. An algorithm visualization of the insertion sort algorithm

Moreover, in students' subsequent presentations, their visualizations tended to stimulate discussions about *implementation details*—for example, how a particular aspect of a visualization was implemented.

These observations led us to change the visualization assignments significantly for the subsequent offering of the course. In particular, inspired by the observational studies of art supplies just reported, we had students use pens, paper, scissors, transparencies, and the like to construct and present "homemade" visualizations that (a) illustrated the target algorithm for a few, carefully-selected input data sets, and (b) tended to have an unpolished, sketched appearance, owing to the fact that they were generated by hand. We found that students spent 6.2 hours on average ($n = 20$) constructing and refining a single homemade visualization. For most of that time, students focused on understanding the target algorithm's procedural behavior, and how they might best communicate it through a visualization. Moreover, rather than stimulating discussions about implementation details, their homemade visualizations tended to mediate discussions about the *underlying algorithm*, and about how the visualizations might bring out its behavior more clearly. Finally, the flexibility of the art supplies enables enabled students to go back and re-present sections of their visualizations, as well as to mark-up and dynamically modify them, in response to audience questions and feedback. As a result, presentations tended to engage the audience more actively in interactive discussions.

These findings, coupled with those of the observational studies reported above, have provided a solid empirical foundation for the design of ALVIS. In ongoing work, we are subjecting ALVIS to iterative usability testing, and refining the semantics of its spatial relations expressions through further observational studies.

Summary And Research Agenda

In this chapter, we have portrayed algorithm visualization as an area of research with two distinct research eras. In the first of these eras, which spanned the 1980s and early 1990s, researchers focused on innovating algorithm visualization technology through, for example, novel displays, novel specification techniques, and novel interaction techniques. In the second of these areas, which began in the mid 1990s and continues to the present, researchers turned their attention to exploring and evaluating the educational benefits of the technology. In order to provide researchers with an overview of, and entry points into, the area, we have reviewed the key contributions and methodologies in each of these eras, and we have presented two case studies of algorithm visualization system design and evaluation.

Our review suggests several directions for future research in the area. Some of the most promising of these include the following:

1. *Greater system interactivity*. Algorithm visualization systems tend to be like movies in that students can view, rewind, and fast forward, but not really interact with the display. Future systems should consider increased interaction capabilities so that viewers can explore "what if" scenarios, grab display objects, and explore algorithm operations.

2. *User-centered technology development.* Developers of algorithm visualization technology tend to be computer science instructors who ultimately use their own technology as teaching aids. As a result, they tend to trust their own intuitions regarding design. While an instructor's perspective is certainly an important ingredient in the design process, we believe that future technology ought to be developed within the context of a learner-centered design process. In such a design process, iterative usability studies with prospective users, rather than designer intuitions, guide the technology's design.

3. *Longitudinal studies of effectiveness.* To date, most experimental studies of effectiveness have focused on learning sessions on the order of a few hours. Given the short length of the learning sessions, it is remarkable that studies have detected significant learning differences. At the same time, one must ask whether the short learning session length typical of these experimental studies has prevented learning outcomes from fully developing. What if, for example, one was to test the benefits of regular, continuous use of algorithm visualization technology over the period of a semester or year? Future experimental studies ought to consider broadening their scope in this way.

4. *Use of evaluation methods other than controlled experimentation.* Despite the scientific allure of controlled experiments and the quantitative results they offer, the mixed results of past controlled experiments reflect the inherent difficulty of asserting causality between learning medium and knowledge acquisition, even in tightly controlled environments. Aside from their potentially low ecological validity, controlled experiments make a potentially "invalid implicit assumption" (Williams & Brown, 1990; see also Payne *et al.*, 1996) by "treat[ing] each medium as a more or less invariant entity with fixed clusters of attributes" (Williams & Brown, 1990, p. 219). Controlled experiments may also encounter difficulties in
    -   controlling for all of the significant variables (Gurka & Citrin, 1996),
    -   manipulating the correct variables (Kehoe, Stasko, & Taylor 2001), and
    -   developing measures that are sensitive to differences in learning promoted by alternative conditions (Kehoe, Stasko, & Taylor 2001).

Researchers should thus think carefully before embarking on controlled experimental investigations of algorithm visualization effectiveness; they may well enjoy more success with other techniques.

References

Badre, A., Baranek, M., Morris, J. M., & Stasko, J. T. (1992). Assessing program visualization systems as instructional aids. In I. Tomek (Ed.), *Computer Assisted Learning, ICCAL '92.* Springer-Verlag, New York, pp. 87-99.

Baecker, R. (1973). Towards animating computer programs: A first progress report, *Proceedings of the Third NRC Man-Computer Communications Conference*, 4.1-4.10.

Baecker, R. (1975). Two systems which produce animated representations of the execution of computer programs. *SIGCSE Bulletin 7*, 158-167.

Baecker, R. (1981). With the assistance of Dave Sherman, Sorting out Sorting, 30 minute color sound film, Dynamic Graphics Project, University of Toronto, 1981. (Excerpted and "reprinted" in SIGGRAPH Video Review 7, 1983. Distributed by Morgan Kaufmann, Publishers.)

Baecker, R., & Price, B. (1998). The early history of software visualization. In J. Stasko, J. Domingue, M. Brown, & B. Price, (Eds.), *Software Visualization: Programming as a Multimedia Experience,* (pp. 29-34). Cambridge, MA: The MIT Press.

Baecker, R. (1998). Sorting out sorting: A case study of software visualization for teaching computer science. In J. Stasko, J. Domingue, M. Brown, & B. Price, (Eds.), *Software Visualization: Programming as a Multimedia Experience,* (pp. 369-381). Cambridge, MA: The MIT Press.

Bazik, J., Tamassia, R., Reiss, S., & van Dam, A. (1998). Software visualization in teaching at Brown University. In J. Stasko, J. Domingue, M. Brown, & B. Price, (Eds.), *Software Visualization: Programming as a Multimedia Experience,* (pp. 383-398). Cambridge, MA: The MIT Press.

Brown M. & Sedgewick R. (1985) Techniques for algorithm animation, *IEEE Software, 2*(1), 28-39.

Brown, M. (1988). Exploring algorithms using Balsa-II, *Computer, 21*(5), 14-36.

Brown M. & Hershberger J. (1992). Color and sound in algorithm animation, *Computer, 25*(12), 52-63.

Brown M. & Najork M. (1993). Algorithm animation using 3D interactive graphics, In *Proceedings of the 1993 ACM Symposium on User Interface Software and Technology* , Atlanta, GA, 93–100.

Byrne, M., Catrambone, R., & Stasko, J. (1999). Evaluating animations as student aids in learning computer algorithms. *Computers & Education, 33*(4), 253-278.

Chaabouni, Z. (1996). *A user-centered design of a visualization language for sorting algorithms.* Unpublished Master's Thesis, University of Oregon, Eugene, OR.

Crescenzi, P., Demetrescu, C., Finocchi, I., & Petreschi, R. (2000). Reversible execution and visualization of programs with LEONARDO. *Journal of Visual Languages and Computing, 11*(2), 125-150.

Demetrescu, C., Finocchi, I., & Stasko, J. (2002). Specifying algorithm visualizations: Interesting events or state mapping?, Proceedings of the International Dagstuhl Seminar on Software Visualization, Schloss Dagstuhl, May 2001, appears in *Software Visualization State-of-the-Art Survey*, LNCS 2269, Stephan Diehl (ed.), Springer Verlag, pp. 16-30.

Douglas, S. A. (1995). Conversation analysis and human-computer interaction design. In P. Thomas (Ed.) In *Social and Interactional Dimensions of Human-Computer Interfaces.* Cambridge University Press, Cambridge.

Douglas, S. A., Hundhausen, C. D., & McKeown, D. (1995). Toward empirically-based software visualization languages. In *Proceedings of the 1995 IEEE Symposium on Visual Languages*. IEEE Computer Society Press, Los Alamitos, pp. 342-349.

Douglas, S. A., Hundhausen, C. D., & McKeown, D. (1996). Exploring human visualization of computer algorithms. In *Proceedings 1996 Graphics Interface Conference*. Canadian Graphics Society, Toronto, pp. 9-16.

Ericsson, K. A., & Simon, H. A. (1984). *Protocol Analysis: Verbal Reports as Data.* MIT Press, Cambridge, MA.

Ford, L. (1993). How programmers visualize programs, *Empirical Studies of Programmers: Fifth Workshop.* Lawrence Erlbaum, Englewood, pp. 224.

Gilmore, D. J. (1990). Methodological issues in the study of programming. In J.-M. Hoc & T. R. G. Green & R. Samurcay & D. J. Gilmore (Eds.), *Psychology of Programming.* Academic Press, San Diego, pp. 83-98.

Gloor, P. (1998). Animated algorithms. In J. Stasko, J. Domingue, M. Brown, & B. Price, (Eds.), *Software Visualization: Programming as a Multimedia Experience,* (pp. 409-416). Cambridge, MA: The MIT Press.

Haajanen, J., Pesonius, M., Sutinen, E., Tarhio, J., Teräsvirta, T., & Vanninen, P. (1997). Animation of user algorithms on the Web. In: *Proceedings of VL '97, IEEE Symposium on Visual Languages*. IEEE Computer Society Press, Los Alamitos, pp. 360-367.

Hix, D., & Hartson, H. R. (1993). *Developing user interfaces: Ensuring usability through product & process*. John Wiley & Sons, New York.

Hundhausen, C. D. (2002). Integrating algorithm visualization technology into an undergraduate algorithms course: Ethnographic studies of a social constructivist approach. *Computers & Education, 39*(3), 237-260.

Hundhausen, C. D., & Douglas, S. A. (2002). Low fidelity algorithm visualization. *Journal of Visual Languages and Computing, 13*(5), 449-470.

Hundhausen, C. D., Douglas, S. A., & Stasko, J. T. (2002). A meta-study of software visualization effectiveness. *Journal of Visual Languages and Computing, 13*(3), 259-290.

Jarc, D. J., Feldman, M. B., & Heller, R. S. (2000). Assessing the benefits of interactive prediction using web-based algorithm animation courseware, *Proceedings SIGCSE 2000*. ACM Press, New York, pp. 377-381.

Jordan, B., & Henderson, A. (1995). Interaction analysis: Foundations and practice. *Journal of the Learning Sciences, 4*(1), 39-103.

Kehoe, C., Stasko, J. T., & Taylor, A. (2001). Rethinking the evaluation of algorithm animations as learning aids: An observational study. *International Journal of Human-Computer Studies, 54*(2), 265-284.

Kraemer, E. & Stasko, J. (1993). The visualization of parallel systems: An overview. *Journal of Parallel and Distributed Computing, 18*(2), 105-117.

Lawrence, A. (1993). *Empirical studies of the value of algorithm animation in algorithm understanding*. Unpublished Ph.D. dissertation, Georgia Institute of Technology, Atlanta.

Miyake, N. (1986). Construction interaction and the iterative process of understanding. *Cognitive Science, 10*, 151-177.

Mulholland, P. (1998). A principled approach to the evaluation of SV: a case study in Prolog. In J. Stasko, J. Domingue, M. Brown, & B. Price, (Eds.), *Software Visualization: Programming as a Multimedia Experience*, (pp. 439-452). Cambridge, MA: The MIT Press.

Mukherjea, S., & Stasko, J. (1994). Toward visual debugging: Integrating algorithm animation capabilities within a source-level debugger. *ACM Transactions on Computer-Human Interaction, 1*(3), 215-244.

Naps, T., Rodger, S., Velazquez-Iturbide, J.A., Rößling, G., Almstrum, V., Dann, W., Fleischer, R., Hundhausen, C., Korhonen, A., Malmi, L., McNally, M. (2003). Exploring the role of visualization and engagement in computer science education. *ACM SIGCSE Bulletin, 35*(2), 131-152.

Pane, J. F., Corbett, A. T., & John, B. E. (1996). Assessing dynamics in computer-based instruction, *Procedings of the 1996 SIGCHI Conference on Human Factors in Computing Systems*. ACM Press, New York, pp. 197-204.

Price, B. (1990). *A framework for the automatic animation of concurrent programs*. Unpublished M.S. Thesis, University of Toronto.

Price, B., Baecker, R., & Small, I. (1993). A principled taxonomy of software visualization. *Journal of Visual Languages and Computing, 4*(3), 211-266.

Roman, G.-C. & Cox, K., (1989). A declarative approach to visualizing concurrent computations, *Computer, 22*(10), 25-36.

Roman, G.-C., Cox, K., Wilcox, C. & Plun J. (1992). Pavane: A system for declarative visualization of concurrent computations, *Journal of Visual Languages and Computing, 3*(1), 161-193.

Rößling, G., and Freisleben, B. (2002). ANIMAL: A system for supporting multiple roles in algorithm animation. *Journal of Visual Languages and Computing, 13*(3), 341-354.

Sanderson, P. M., & Fisher, C. (1994). Exploratory sequential data analysis: Foundations. *Human-Computer Interaction, 9*(3-4), 251-318.

Sanjek, R. (1995). Ethnography. In A. Barnard & J. Spencer (Eds.), *Encyclopedic dictionary of social and cultural anthropology*. Raitledge, London.

Spradley, J. P. (1979). *The Ethnographic Interview*. Holt, Rinehart, and Winston, New York.

Stasko, J. (1990a) TANGO: A framework and system for algorithm animation. *Computer, 23*(9), 27-39.

Stasko, J., (1990b). The path-transition paradigm: A practical methodology for adding animation to program interfaces," *Journal of Visual Languages and Computing, 1*(3), 213-236.

Stasko, J. (1991). Using direct manipulation to build algorithm animations by demonstration. *Proceedings of ACM CHI'91 Conference on Human Factors in Computing Systems*, 307-314.

Stasko, J., Badre, A. & Lewis, C. (1993). Do algorithm animations assist learning? An empirical study and analysis. *Proceedings of ACM INTERCHI '93 Conference on Human Factors in Computing Systems*, 61-66.

Stasko J. & Kraemer E., (1993). A methodology for building application-specific visualizations of parallel programs, *Journal of Parallel and Distributed Computing, 18*(2), 258-264.

Stasko J. and Wehrli J., (1993). Three-dimensional computation visualization, *Proceedings of the 1993 IEEE Symposium on Visual Languages*, Bergen, Norway, 100-107.

Stasko, J. (1997). Using student-built animations as learning aids. *Proceedings of the ACM Technical Symposium on Computer Science Education*, 25-29.

Stasko, J., Domingue, J., Brown, M. & Price, B. (Eds.) (1998). *Software Visualization: Programming as a Multimedia Experience*. Cambridge, MA: MIT Press.

Stasko, J. (1998a). Smooth, continuous animation for portraying algorithms and processes. In J. Stasko, J. Domingue, M. Brown, & B. Price, (Eds.), *Software Visualization: Programming as a Multimedia Experience,* (pp. 103-118). Cambridge, MA: The MIT Press.

Stasko, J. (1998b). Building software visualizations through direct manipulation and demonstration. In J. Stasko, J. Domingue, M. Brown, & B. Price, (Eds.), *Software Visualization: Programming as a Multimedia Experience,* (pp. 187-203). Cambridge, MA: The MIT Press.

Williams, C. J., & Brown, W. W. (1990). A review of research issues in the use of computer-related technologies for instruction: What do we know? *Journal of Instructional Media, 17*(3), 213-225.

Wolcott, H. F. (1992). Posturing in qualitative inquiry. In M. D. LeCompte & W. L. Millroy & J. Preissle (Eds.), *The Handbook of Qualitative Research in Education*. Academic Press, San Diego, pp. 3-52.

Notes

1 See (Gilmore, 1990) for a more comprehensive treatment of observational techniques.

2 Some might contest our choice not to include interviews in this category. Because of the interview's prominence and heritage as an ethnographic field technique (see Spradley, 1979), we include it in that category, with the caveat that this classification may not be agreeable to everyone.

3 The term ethnography has a double meaning, being both a collection of research techniques, and a genre of reportage (Sanjek, 1995). Thus, our decision to use the term ethnographic field techniques, as opposed to ethnography, is quite deliberate.

4 See our justification for classifying interviews as an ethnographic field technique in footnote 2. See Spradley (1979) for an overview of the many ethnographic interviewing techniques, including structured, semi-structured, and casual.

5 See (Hix & Hartson, 1993) for a more extensive treatment.

6 The technique of having participants work in pairs is known as constructive interaction (Miyake, 1986); Douglas (1995) considers its advantages over single-participant usability studies.

7 We use both Behaviorist tradition and Cognitivist tradition strictly in the senses in which they are used in (Sanderson & Fischer, 1994).

8 However, note that, in order to create a more natural social situation, some researchers opt to employ groups of participants in usability studies; see Footnote 6.

List of Contributors

Sally Fincher
Computing Laboratory
University of Kent
Canterbury
UK

Michael Clancy
The Computer Science Division
University of California, Berkeley
Berkeley, CA
USA

Mark Guzdial
College of Computing
Georgia Institute of Technology
Atlanta, GA
USA

W. Michael McCracken
College of Computing
Georgia Institute of Technology
Atlanta, GA
USA

John T. Stasko
College of Computing
Georgia Institute of Technology
Atlanta, GA
USA

Marian Petre
Faculty of Mathematics and Computing
Open University
Milton Keynes
UK

Tony Clear
School of Computer and Information
Science
Auckland University of Technology
Auckland
New Zealand

Christopher D. Hundhausen
School of Electrical Engineering and
Computer Science
Washington State University
Pullman, WA
USA

Robert S. Rist
Faculty of Information Technology
University of Technology, Sydney
Australia

Index

Note: Page numbers in italics indicate figures or tables; a page number followed by "n" and a number indicates an endnote and note number.

Printed and bound by CPI Group (UK) Ltd, Croydon, CR0 4YY

24/10/2024

01778295-0002